3824

Adaptive Phased Array
Thermotherapy for Cancer

For a listing of related titles
turn to the back of this book.

Adaptive Phased Array Thermotherapy for Cancer

Adaptive Phased Array Thermotherapy for Cancer

Alan J. Fenn

ARTECH
HOUSE

BOSTON | LONDON
artechhouse.com

Library of Congress Cataloging-in-Publication Data
A catalog record for this book is available from the Library of Congress.

British Library Cataloguing in Publication Data
A catalogue record for this book is available from the British Library.

ISBN-13: 978-1-59693-379-8

Cover design by Igor Valdman

This work was sponsored under Air Force Contract FA8721-05-C-0002. Opinions, interpretations, conclusions, and recommendations are those of the author and are not necessarily endorsed by the United States Government.

10 9 8 7 6 5 4 3 2 1

To My Family

Contents

Preface

Common approaches for treating cancer include surgery, radiation therapy, chemotherapy, and hormonal therapy; however, these approaches do not always completely eliminate or kill all of the cancer cells and there are many side effects. New treatment approaches, typically used in combination with surgery, radiation therapy, chemotherapy, and hormonal therapy, are desirable that can increase tumor cell kill without added side effects. Elevated temperature (hyperthermia or thermotherapy) is known to kill cancer cells and it is synergistic with radiation therapy and chemotherapy. Heat can be delivered to tissues, particularly for high-water high-ion content tumors, by radiofrequency or microwave radiation from array antennas. However, with external radiofrequency or microwave array antenna applicators, without proper control it can be difficult to heat tumors in the body without burning the superficial or deep healthy tissues.

Adaptive microwave phased array antennas are well known for their ability to improve the performance of communications and radar systems operating in the presence of interference. Adaptive phased array techniques have recently been applied to radiofrequency and microwave thermotherapy treatment of cancerous tumors. This monograph is written primarily for graduate students in biomedical engineering, electrical engineering, and medical physics, and for cancer researchers and oncologists, and describes research on adaptive phased array antenna techniques for cancer treatment developed at the Massachusetts Institute of Technology, Lincoln Laboratory. Some of the material in this book has been published previously only in MIT Lincoln Laboratory reports, and some of the material has been published in the peer-reviewed open literature. Some background in signal processing, electromagnetic theory, antennas, medical physics, and oncology will be helpful to the reader.

Chapter 1 begins with background on cancer treatments using thermotherapy and introduces the concept of adaptive radiofrequency and microwave phased array treatment of cancer. Adaptive phased array algorithms for thermotherapy, including the sample matrix inversion algorithm and the gradient-search algorithm are described in Chapter 2. Chapter 3 reviews electromagnetic theory for waves in conducting media in the context of using electromagnetic energy to heat body tissues. Chapter 4 discusses thermal modeling theory that is used to compute the thermal distribution in a target body that is heated using electromagnetic radiation fields as the driving source. Computer simulations for adaptive phased arrays heating the deep torso are described in Chapter 5. Chapter 6 describes phantom measurements for an adaptive phased array heating the deep torso using multiple types of phantoms. Chapter 7 describes the design and finite-difference time-domain analysis of a monopole phased array thermotherapy applicator for deep heating. Chapter 8 reviews preclinical results for an adaptive phased array focused microwave thermotherapy system for treating tumors in the intact breast. These preclinical results include finite-difference time-domain analysis, phantom measurements, and animal testing. Clinical results for an adaptive phased array focused microwave thermotherapy system used in treating invasive breast cancer in the intact breast are presented in Chapter 9. Future research topics on adaptive phased array thermotherapy are outlined in Chapter 10.

The support of the MIT Lincoln Laboratory Advanced Concepts Committee for the initial proof-of-concept analysis and phantom measurements described in Chapter 6 is sincerely appreciated. Significant support and encouragement for development and demonstration of the adaptive phased array thermotherapy technology at the MIT Lincoln Laboratory came from, most notably for early preclinical development, Mr. Donald H. Temme, Mr. James Fitzgerald, and Mr. Robert J. Burns. The author is grateful to Ms. Lori Pressman of the MIT Technology Licensing Office for providing technology transfer support. The author is also grateful to Madeline Riley of the MIT Lincoln Laboratory for coordinating the preparation of the graphics.

For the measured preclinical phantom testing results presented in Chapter 6, the author is grateful to P.F. Turner for technical discussions and to J. Robbins for software support, both of BSD Medical Corporation. For this preclinical phantom testing, sincere gratitude is extended to both Dr. Gerald A. King of the State University of New York, Health Science Center in Syracuse and to Dr. V. Sathiaseelan of Northwestern Memorial Hospital in Chicago.

The author is grateful to Dr. Allen Taflove and Christopher E. Reuter at Northwestern University for supporting the finite-difference time-domain computer simulations presented in Chapters 7 and 8.

For the animal trials results presented in Chapter 8, the author is grateful to Dr. Jeffrey W. Hand of Hammersmith Hospital in London, UK, for coordinating and conducting the animal studies. The author also expresses gratitude to Dr. Augustine Y. Cheung, Mr. John Mon, and Mr. Dennis Smith of Celsion (Canada) Limited for developing the clinical breast thermotherapy system described in Chapters 8 and 9. In Chapter 9, the data for the breast cancer patients treated with the adaptive phased array focused microwave thermotherapy system are results of the efforts of a number of colleagues, most notably for the clinical results: Dr. Hernan I. Vargas, Dr. William C. Dooley, Dr. Robert A. Gardner, Dr. Sylvia Heywang-Köbrunner, Dr. Mary Beth Tomaselli, Dr. Jay K. Harness, Dr. Christine T. Mroz, Dr. Lynne P. Clark, Dr. Claire M. Carman, Dr. Sandra B. Schultz, Dr. John Winstanley, Dr. Gary V. Kuehl, Dr. Mariana Doval, and Dr. Jerome B. Block, and their efforts are sincerely appreciated.

1

Adaptive Phased Array Thermotherapy Technique

1.1 INTRODUCTION

An antenna is a transducer that can convert both an incident electromagnetic wave to a signal voltage on a transmission line and a signal voltage on a transmission line to a transmitted electromagnetic wave [1-3]. As depicted in Figure 1.1, a phased array antenna consists of a minimum of two antenna elements that typically transmit or receive signals with a proper timing (phase relation) between the elements and perform amplification such that the summed signal of the elements has a desired maximum value in a particular direction [4-20]. Adaptive phased array antennas can be used in radar systems to adjust the relative phase and amplitude weighting of the array elements to null or reduce the effects of interference (jamming) and to improve the received signal-to-noise ratio (SNR) of desired target signals under far-field conditions [20-34] and under focused near-field testing conditions [20, 35-40]. Electromagnetic transmitting antennas and phased arrays can be used clinically in heating and killing cancerous tumors in the body [41-81]. In this book, adaptive electromagnetic transmit phased array antennas are explored in the application of adaptive phased array heat treatment (thermotherapy) for cancer [82-98].

Consider Figure 1.2, which shows the general case of M interference sources in the field of view of an N-element fully adaptive receive array antenna for radar or communications applications. A fully adaptive array antenna uses all of the antenna element channels in forming adaptive nulls while maintaining gain in a specified direction – a partially adaptive array uses only some of the channels to form nulls. The N-element fully adaptive

1

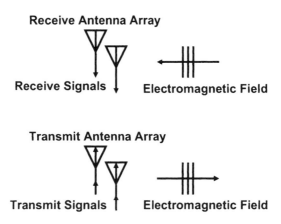

Figure 1.1 Simplified diagram for two-element receive and transmit antenna arrays.

array has $N - 1$ distinct nulls available to adaptively suppress interferers [99-104]. If there are $M = N - 1$ interference sources distributed within the N-channel fully adaptive antenna field of view, they can be nulled completely. The time-dependent N complex weights, denoted $w_1(t), w_2(t), \cdots, w_N(t)$, are adaptively controlled in amplitude and phase to null the interference characterized by the received signals $S_1(t), S_2(t), \cdots, S_N(t)$, so that the summed output voltage, $y(t)$, has a reduced amount of interference. After adaption, the SNR achieved depends on the null depths achieved and the amount of reduction in radiation pattern gain (if any) to a desired direction of a radar target. For radar or communications applications, adaptive phased arrays are normally used in a receiving mode as depicted in Figure 1.2. In the case of cancer treatment, the adaptive phased array is used in a transmit mode with sufficient transmit power to heat and kill cancer cells, while avoiding damage to surrounding normal healthy body tissues and organs.

Radar systems normally operate with targets that are located many kilometers from the radar antenna [105, 106]. However, an adaptive phased array radar system can be tested close to the antenna aperture by using a focused near-field adaptive phased array approach investigated by the author [35-40]. The phase shifters in a phased array antenna can be used to focus the array in the near-field such that the near-field radiation pattern is approximately equal to the far-field radiation pattern [107]. Figure 1.3 shows the general concept of a focused near-field adaptive nulling array antenna. Here, a focal point and a number of adaptive nulls are located at a range distance, denoted r, of about one to two aperture diameters from the adaptive phased array antenna. Analysis has shown that an adaptive

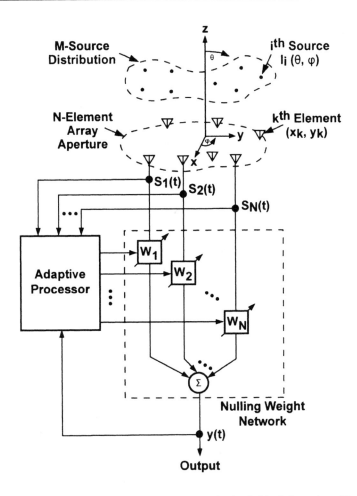

Figure 1.2 Distribution of M interference sources in the field of view of an N-element adaptive nulling receive array antenna for radar or communications applications. (© 1985 IEEE [102].)

phased array antenna radar system can be tested in the focused near-field region with performance comparable to radar operation at normal range distances to the target and interference [39]. In Figure 1.4, the contrast between adaptive receive phased array radar system testing in the focused near-field region and cancer treatment using a focused near-field adaptive nulling transmit phased array antenna is depicted. For adaptive phased array radar testing, the array is focused in the direction of target sources and then radar clutter and jammer sources are distributed in the field of view [39]. For cancer treatment, the array focus is at the tumor target and adaptive nulls are aimed at normal healthy tissues to avoid side effects (burning).

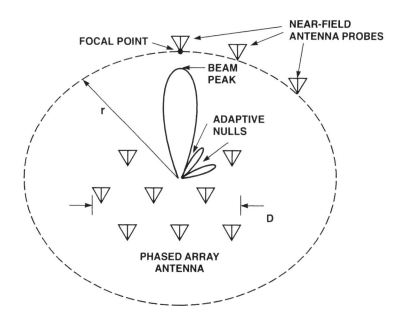

Figure 1.3 General concept for focused near-field adaptive nulling.

For cancer treatment, the tumor target is located typically less than one-half meter from the transmitting antenna applicator, with the intervening tissue presenting a significant amount of signal attenuation due to dielectric losses. The variation in tissue parameters such as dielectric constant, specific heat, thermal conduction, tissue density, blood flow, and metabolic rate [108], from patient-to-patient or from treatment-to-treatment, presents a significant challenge for reliable heating of deep tumors.

In the focused near-field adaptive nulling technique described here, it is assumed that the hyperthermia phased array antenna is focused in the near field and that a main beam and, possibly, sidelobes or large amplitude fields are formed in the target tissue. It is assumed that phase focusing is used to produce the desired main beam before adaptive nulling is effected. Figure 1.5 shows a probe antenna, located at a desired focal point of an array, that is used in phase calibration. The array can maximize the signal received by the calibration probe by adjusting its phase shifters adaptively so that the observed element-to-element phase variation is effectively removed. The resulting near-field radiation pattern will have a main beam and sidelobes. The main beam will be pointed at the array focal point, and sidelobes will exist at angles away from the main beam. Auxiliary probe antennas can be placed at the desired null positions in the quiescent (before nulling) sidelobe region or regions of

RADAR SYSTEM TESTING　　　　　　　**HYPERTHERMIA**

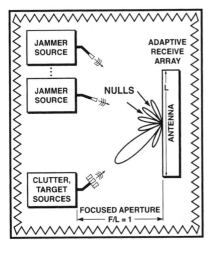
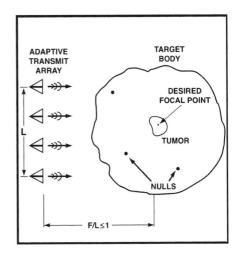

ANECHOIC CHAMBER

Figure 1.4 Two applications of focused near-field adaptive nulling. Left: radar system testing. Right: hyperthermia treatment of cancer.

large amplitude fields. In the context of cancer treatment, these sidelobes or large amplitude fields are where tissue hot spots are likely to occur; they are nulled by an adaptive nulling system. The main problems in hyperthermia treatment of cancer are the formation of hot spots at tissue interfaces and skin burns.

As an example of focused near-field adaptive nulling in a radar antenna context, consider the radiation pattern characteristics of a 32-element linear array of isotropic elements with one-half wavelength spacing operating in free space at the microwave frequency 1.3 GHz. This linear array example has an aperture length of 3.58m. Let L be the maximum length of the aperture, let F be the focal distance of the aperture, and let λ be the wavelength. The conventional approximate far-field test distance $(2L^2/\lambda)$ for an aperture this size is about 110m; however, the array can be focused in the near field to achieve radiation pattern characteristics approximately equal to the far-field characteristics. Let the array be focused at a distance of one aperture diameter $(F/L = 1)$ at the angle $60°$ with respect to the axis of the linear array. Thus, the focal range for this example is 3.58m. A two-dimensional gain radiation contour pattern for $F/L = 1$ for this antenna array is shown in Figure 1.6(a) [37, Fenn, 1990]. The gain pattern contours are expressed in decibels relative to an isotropic (omnidirectional) receive antenna element. A beam maximum is located at the position of the desired focal point. Next, let an interference source be located on a radiation pattern sidelobe 33° from the axis of the array, also at one aperture diameter distance. The array weights are adjusted by the sample matrix inversion algorithm as described in Chapter 2. The two-dimensional gain radiation pattern for the adaptive array is given in Figure 1.6(b). A radiation pattern null is located, as desired, at the interference source position, while the beam maximum remains located at the focal point. Referring to Figure 1.4, to visualize adaptive phased array hyperthermia treatment of cancer, the near-field focused radiation maximum would be located at a tumor and the adaptive radiation pattern nulls would be located in or near healthy tissue or organs.

This book is organized as follows. The next two sections discuss background information for cancer and for hyperthermia treatment of cancer. The subsequent section in this chapter describes conceptual details of adaptive phased array thermotherapy treatment of cancer. Chapter 2 describes two types of adaptive phased array algorithms for cancer treatment, the sample matrix inversion (SMI) algorithm, and the gradient search (GS) algorithm – these types of algorithms are the same as used by radar and communications systems. Theory for analyzing the electromagnetic fields generated by an adaptive phased array thermotherapy system is given in Chapter 3. In Chapter 4, theory for computing the thermal distribution induced by electromagnetic fields in tissue is given. Simulated electromagnetic fields and deep-heating thermal distributions for an adaptive phased array are presented in Chapter 5. Next, in Chapter 6 preclinical measurements

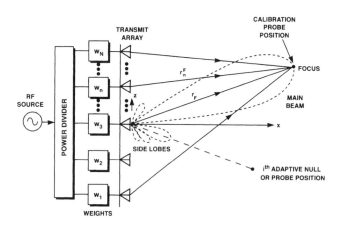

Figure 1.5 Adaptive transmit phased array antenna near-field focusing concept.

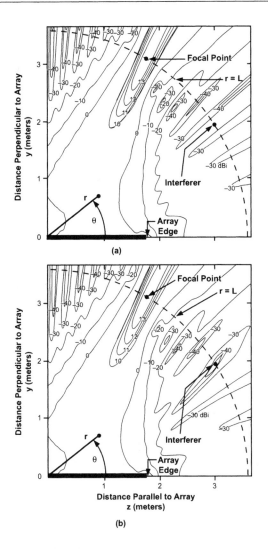

Figure 1.6 Calculated two-dimensional near-field gain radiation contour patterns for a near-field focused 32-element linear array (one-half of the array is shown as a thick black bar). A single interferer is at range $F/L = 1$ ($r_i = 3.58$m) and angle $\theta_i = 33°$. (a) Before adaptive nulling and (b) after adaptive nulling. (© 1990 IEEE [37].)

in phantom tissue representing a torso are presented for a dipole phased array hyperthermia system modified to implement adaptive phased array gradient-search algorithms. The preclinical measurements are performed in a homogeneous saline-filled cylindrical phantom, a heterogeneous beef phantom, and a homogeneous saline-filled light-emitting diode elliptical phantom. In Chapter 7, design, analysis, and phantom test results for a monopole phased array hyperthermia applicator are described. Chapter 8 describes preclinical investigations of adaptive phased arrays for treatment of breast cancer. In Chapter 9, clinical results for adaptive phased array thermotherapy treatment of breast cancer in the intact breast are described. Chapter 10 discusses future research areas for adaptive phased array thermotherapy.

1.2 BACKGROUND FOR CANCER TREATMENT

Solid cancerous tumors can be described as a growing mass of malignant tumor cells that usually invade, interfere with, and damage normal organs in the body. Cancers can develop as primary local tumors that can spread (metastasize) to other regions and organs of the body. Tumors that have spread from one organ to another are referred to as secondary cancers. Approximately 1.5 million new cases of invasive cancer were expected to be diagnosed in 2007 in the United States [109]. A summary of some of the expected number of new cases of common cancers in the United States is given in Table 1.1. In addition to the cancers listed in Table 1.1, more than 1 million new cases of skin cancer were expected in 2007 in the United States.

According to World Health Organization analysis, cancer and heart disease are currently the two leading causes of death worldwide [110]. Cancer

Table 1.1
Estimated New Cases of Invasive Cancer in 2007 (United States) [109]

Cancer Site	Number of New Cases (Both Sexes)
Bladder	67,160
Brain	20,500
Breast	180,510
Colon	112,340
Esophagus	15,560
Kidney and renal pelvis	51,190
Liver	19,160
Lung and bronchus	213,380
Ovary	22,430
Pancreas	37,170
Prostate	218,890
Rectum	41,420
Stomach	21,260
Uterine	50,230

primarily affects the elderly, and the elderly population is rising annually. Globally, by the year 2030, cancer deaths are expected to rise by about 20% compared to the annual deaths in 2009. With the annually increasing burden of cancer worldwide [111, 112], improvements in cancer treatments are increasingly desirable.

The successful treatment of deep-seated malignant tumors, such as those listed in Table 1.1, is often a difficult task. The objective of the treatment is to destroy, reduce in size, or completely remove the tumor mass by one or more modalities available at the cancer treatment facility. Common treatment modalities are surgery, radiation therapy, chemotherapy, and hormonal therapy used alone or in combination with one another according to patients' specific needs [113-126]. While surgery, radiation therapy, chemotherapy, and hormonal therapy have some effectiveness in cancer treatments, it is difficult to eliminate all of the cancer cells; thus, there is significant room for improvement in these treatments.

Heat treatment (referred to in this book as hyperthermia or ther-motherapy) is known to kill cancer cells [127-156] and improve tumor response as demonstrated in a number of clinical studies in the last 20 years [149-156]. The next section describes the background for hyperthermia treatment of cancer. Cancer cells can be heated and ablated by a variety of energy sources including lasers, ultrasound, radiofrequency waves, and microwaves [157-162]. This book investigates minimally invasive radiofre-quency and microwave hyperthermia treatment of deep cancers. In the field of hyperthermia, radiofrequency hyperthermia typically is used to refer to frequencies below about 300 MHz, and microwave hyperthermia typically refers to frequencies above about 300 MHz. Typical microwave hyperthermia frequencies are at 433 MHz, 915 MHz, and 2,450 MHz [43, p. 154, 176]. The next two sections review the background for hyperthermia in the treatment of cancer and methods for applying hyperthermia treatments.

1.3 BACKGROUND FOR THERMOTHERAPY TREATMENT OF CANCER

Tissue heating, or hyperthermia [43, 127-156], is a modality used alone or in conjunction with surgery, radiation therapy, and chemotherapy as described in this section. Normal body temperature is $37°C$, and therapeutic hyperthermia for cancer refers to a targeted elevated tumor temperature. When tumor or healthy tissue is elevated sufficiently in temperature for a sufficient period of time, it can be killed [145-148]. Elevated cell temperature can produce severe heat shock that results in protein denaturization followed by inactivation of protein synthesis, inhibition of cell cycle progression, and inhibition of DNA repair processes such that the cells are killed or they are sensitized to radiation therapy or chemotherapy, allowing radiation therapy and/or chemotherapy to become more effective.

A cumulative or equivalent thermal dose is often used to quantify the thermal dose given during thermotherapy treatments. The cumulative or total equivalent thermal dose relative to 43°C is calculated as [147]

$$\text{CEM}(43°\text{C}) = \Delta t \sum_{i=1}^{N} R^{(43-T_i)} \qquad (1.1)$$

where $\sum$ is the summation over a series of N temperature measurements during the treatment, T_i is the series of temperature measurements $(T_1, T_2, \cdots, T_N)$, Δt is the constant interval of time (units of minutes, usually a fraction of a minute) between temperature measurements, R is a rate equal to 0.5 if $T_i \geq 43°\text{C}$ and R is equal to 0.25 if $T_i < 43°\text{C}$. The calculation of cumulative equivalent minutes thermal dose is useful for correlating with tumor cell kill as well as any possible heat damage to tissues including cancerous tissues, healthy skin, and other tissues. Equation (1.1) is a theoretical model developed by Sapareto and Dewey [147] based on extensive in vitro and in vivo cell survival data, and the use of 43°C for the reference temperature is a best estimate for when thermotherapy begins to cause a faster rate of cancer cell kill. Table 1.2 summarizes the equivalent thermal dose in one minute of constant heating as the temperature is varied from 40° to 54°C. As an example in the use of (1.1), if the tissue temperature is maintained at 44°C for 15 minutes, the cumulative equivalent minutes thermal dose is calculated to be CEM(43°C) = $15 \times 2^{(45-43)} = 15 \times 4 = 60$ minutes. An equivalent thermal dose of 15 to 60 minutes relative to 43°C is often sufficient to achieve a therapeutic effect when combined with radiation therapy or chemotherapy. Sufficient energy must be applied to induce the desired temperature rise in tissue.

Electrical energy consumption is commonly expressed in units of kilowatt hours. Mathematically, the expression for the radiofrequency energy W delivered by an applicator can be expressed as [163]:

$$W = \Delta t \sum_{i=1}^{N} P_i \qquad (1.2)$$

In (1.2), Δt represents the constant intervals (in seconds) in which radiofrequency power is applied and the summation is over the complete treatment interval with the power (in Watts) in the ith interval denoted by P_i. The radiofrequency energy W has units of watt-seconds, which is also designated as joules. For example, in three consecutive 60-second intervals if the radiofrequency power is 500 watts, 400 watts, and 600 watts, respectively, the total microwave energy dose delivered in 180 seconds is calculated as $W = 60 \times (500 + 400 + 600) = 90,000$ watt-seconds $= 90,000$ joules or 90 kilojoules. A typical radiofrequency thermotherapy treatment for deep heating with a phased array applicator would use on the order of 1000 watts for a period of about 1800 seconds (30 minutes) which is equal to 1,800,000 joules or 1.8 megajoules.

As shown in Table 1.3, a number of clinical studies of hyperthermia have been published over the last 20 years (since 1988) in which thermotherapy was added to radiation therapy, and improvements in both tumor complete clinical response and patient survival were observed compared to radiation therapy treatment alone [149-155]. In Table 1.3, clinically improved results using hyperthermia and radiation were obtained for treatment of neck nodes, chest wall cancer (recurrent breast cancer), melanoma (skin cancer), brain cancer, bladder cancer, cervical cancer (two studies), and superficial cancers. The results shown in Table 1.3 are statistically significant since the P-values (probability values) are less than 0.05. Taking an unweighted mean value

Table 1.2

Calculated Equivalent Thermal Dose Factor CEM(43°C) from (1.1) for 1 Minute of Constant Heating as the Temperature is Varied from 40° to 54°C

Temperature, T (°C)	Difference $(T - 43°)$	Exponential Factor $R^{(43° - T)}$	Equivalent Thermal Dose Factor (minutes/minute)
40	−3	4^{-3}	0.0156
41	−2	4^{-2}	0.0625
42	−1	4^{-1}	0.250
43	0	2^0	1
44	1	2^1	2
45	2	2^2	4
46	3	2^3	8
47	4	2^4	16
48	5	2^5	32
49	6	2^6	64
50	7	2^7	128
51	8	2^8	256
52	9	2^9	512
53	10	2^{10}	1024
54	11	2^{11}	2048

$R = 0.5$ if $T_i \geq 43°C$, and $R = 0.25$ if $T_i < 43°C$.

of the complete tumor response data summarized in Table 1.3 results in an average complete tumor response rate of 67.1% for hyperthermia plus radiation therapy versus 41.5% for radiation therapy alone. Thus, hyperthermia combined with radiation therapy significantly improves complete tumor response compared to radiation therapy alone. Furthermore, from Table 1.3 overall survival was improved when hyperthermia was administered to patients with cancers in the neck, brain, and cervix, all with statistical significance. In these studies, mild temperature hyperthermia is used in combination with radiation therapy to treat cancer at temperatures in the range of about 39° to 43°C.

The National Comprehensive Cancer Network (NCCN) guidelines for invasive breast cancer has recently been updated to include an option for hyperthermia treatment as described by Carlson [125]. For Stage IV metastatic or recurrent breast cancer in a localized clinical scenario, the current NCCN guideline adds the consideration of combining hyperthermia with radiation therapy as investigated by Jones [154]. For superficial tumors, in a randomized study of 122 enrolled patients, the complete tumor response rate was 66.1%

Table 1.3
Summary of Randomized Clinical Studies Published Since 1988 Showing
Statistically Significant Positive Results for Thermotherapy (TT) Added to Radiation
Therapy (RT) Compared to Radiation Therapy Alone [149-155]

Study	Pub. Date	No. Patients	Cancer	End Point	Results RT+TT	Results RT	P
Valdagni [150]	1988	41	Neck nodes	CR rate OS 5 yrs.	83% 53%	41% 0%	0.016 0.02
Vernon [149]	1996	306	Chest wall	CR rate	59%	41%	<0.001
Overgaard [151]	1996	70 (128 tumors)	Mela-noma	CR rate LC 2 yrs.	62% 46%	35% 28%	0.003 0.008
Sneed [153]	1998	68 (evalu-able)	Brain (GBM)	OS 2 yrs.	31%	15%	0.02
van der Zee [152]	2000	101 114	Bladder Cervix	CR rate CR rate OS 3 yrs.	73% 83% 51%	51% 57% 27%	0.026 0.003 0.009
Harima [155]	2001	40	Cervix	CR rate	80%	50%	0.048
Jones [154]	2005	108	Super-ficial	CR rate	66%	42%	0.02

CR, complete response; LC, local control; OS, overall survival; P≤0.05 significant.

in the hyperthermia combined with radiation therapy arm ($n = 56$), compared to 42.3% in the radiation therapy alone arm ($n = 52$) ($P = 0.02$). In this study, patients had superficial tumors in the breast/chest wall, head and neck, melanoma, or other superficial tumors. Hyperthermia was administered using 433-MHz microwave applicators and delivered to heatable tumors twice a week for up to 10 hyperthermia treatments total, each 1 to 2 hours in duration. The hyperthermia treatments were spaced by a minimum of 48 hours and the targeted equivalent thermal dose (refer to (1.1)) was greater than 10 minutes relative to 43°C.

Cancerous tumors tend to have a damaged vasculature that reduces blood flow into the tumor, and causes them to have a reduced oxygen content. This decreased oxygenation condition is referred to as hypoxia [164], and these hypoxic tumors tend to be radiation-resistant as described by Hahn [132, pp. 32, 33] and Zaffaroni [165]. These hypoxic tumors tend to be chemotherapy-resistant as well [166]. Additionally, damaged tumor vasculature can impede the delivery of systemic chemotherapy agents to tumors as discussed by Jain [167]. When tumors are heated using thermotherapy, blood flow into the tumor can increase, which increases the oxygen content of the tumor allowing radiation therapy to be more effective as described by Song [168]. Thus, thermotherapy should be administered before radiation therapy to utilize the increased tumor oxygenation that occurs on the day of or the day after tumor heating [169, 170]. Similarly, heating the tumor and increasing the blood flow into the tumor can increase the amount of infused chemotherapy that reaches the tumor.

1.4 BACKGROUND FOR DELIVERING THERMOTHERAPY TO DEEP CANCERS

Hyperthermia, or thermotherapy, can be considered as a form of high fever within the body; a controlled thermal dose distribution is required for hyperthermia to have a therapeutic value for tumors without added side effects such as burns. Hyperthermia is typically administered for local control of tumors, to shrink tumors and prevent local cancer recurrence. For some cancers, hyperthermia can improve patient survival as was discussed in the previous section. Hyperthermia can be delivered as a local, regional, or whole body treatment using various technologies [157]. For radiofrequency (RF) and microwave thermotherapy, heating is due to molecular friction of polarized water molecules that rotate and collide in response to a time-varying electromagnetic field in tissue. In the case of deep-regional hyperthermia, heating is accomplished using RF electromagnetic fields in the range of about 8 MHz to 150 MHz. Capacitive plate RF electrodes [156] and electromagnetic phased arrays [41-81] have been investigated for delivering heat to desired regions of the body.

Local treatment refers to treatment of just the tumor tissue. Regional treatment refers to treatment of the tumor tissue and a region of tissue surrounding the tumor. Whole-body hyperthermia is administered to patients with cancer that has spread in the body (metastatic cancer). Regional treatments are used in treating an entire organ such as the liver, stomach, cervix, bladder, prostate, or breast. Typical localized-hyperthermia temperatures required for therapeutic treatment of cancer are in the range of about 39° to 45°C when used in combination with radiation therapy or chemotherapy. For heat-alone thermotherapy treatment to ablate (kill) cancerous tumors, temperatures in the range of at least 48° to 52°C are necessary [96]. Surgical tumor tissue ablation with high-intensity focused ultrasound, interstitial RF, or interstitial laser therapy involves temperatures above about 60°C [158-162].

Based on clinical trials of adaptive phased array thermotherapy with external (transcutaneous) applicators [95-98, 174] for breast cancer, the estimated achievable therapeutic temperature range is about 39° to 52°C as depicted in Figure 1.7. For treatment of breast cancer in the intact breast, the therapeutic range of about 39° to 46°C would be used in combination with chemotherapy and the range of 48° to 52°C would be used as a preoperative heat-alone ablative treatment. Theoretically, the adaptive phased array thermotherapy system is capable of clinically treating many different types of deep-seated tumors (cancerous and benign) such as those occurring

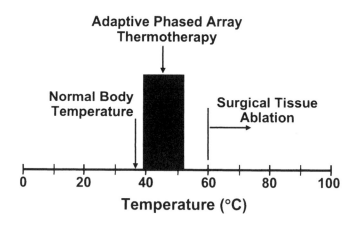

Figure 1.7 Temperature scale showing the approximate range of tumor treatment temperatures used in adaptive phased array thermotherapy treatments to date [95-98, 174] and used in surgical tissue ablation.

in the prostate, breast, liver, rectum, colon, cervix, pancreas, stomach, bladder, lung, and other deep organ sites in the human body. This type of thermotherapy system can be used to target the delivery of drugs by heating the tissue and releasing drugs from thermosensitive liposomes circulating within the bloodstream in the vicinity of the targeted tissue. The same thermotherapy system can also be used in combination with targeted radiation therapy and gene therapy. In contrast with photodynamic therapy, which uses laser light to energize drugs, deep heating with a noninvasive adaptive phased array thermotherapy system may be used to activate thermosensitive liposomes to concentrate a drug into a tumor and energize the drug. Normal tissue should be kept at temperatures below 43°C during the treatment to avoid possible heat-damaging side effects. Various energy sources can be used to induce an elevated temperature in tumor tissue, including radiofrequency, microwave, ultrasound, and lasers. These energy sources can be applied either internally (interstitially [141]) or externally (transcutaneously [142]) to the target tissue. Intracavitary applicators [157] can be used in treating tumors located within or adjacent to body cavities as in the rectum, urethra, bladder, vagina, cervix, uterus, or esophagus. The most difficult aspect of implementing thermotherapy externally, with either radiofrequency (RF) waves, microwaves (MW), or acoustic (ultrasound (US)) waves, is producing sufficient heating at depth without burning the skin. Multiple-applicator RF, MW, or US hyperthermia arrays can be used to provide a focused radiation beam at the tumor position. In this book, the phrase *electromagnetic heating* will refer to RF and MW hyperthermia.

An early conceptual development of a deep-heating electromagnetic phased array hyperthermia applicator was published in a Massachusetts Institute of Technology report in 1973 by von Hippel, Runck, and Westphal [171, p. 17]. Figure 1.8, reproduced from the report, shows a four-horn (rectangular waveguide) antenna array applicator surrounding the torso. The report described the use of a bolus of fluid to couple energy from the radiofrequency applicator array into the torso. The electric field of the four-horn antenna array was aligned with the axis of the torso and allows summing the electric fields for focused heating. Figure 1.9 is redrawn from the von Hippel [171, p. 19] report and shows an opposing phased array antenna applicator in which a power divider provides two channels for simultaneous operation, and a phase shifter is used to provide a relative phase shift between the two channels to focus (maximize) radiofrequency energy between the applicators for heat treatment of a tumor. Subsequently, in 1979 Anderson and Melek [62] described a focused microwave array for heating tumors in body tissues and then numerous research articles on phased array hyperthermia followed [43-61, 63-81].

For localized treatments, a focal region should be concentrated at the tumor with minimal energy delivered to surrounding normal tissue. As the thermotherapy antenna beamwidth is proportional to the wavelength, a small focal region suggests that the RF wavelength be as small as possible. However, due to propagation losses in tissue, the RF depth of penetration decreases with increasing transmit frequency. One of the major problems in heating a deep-seated tumor with a hyperthermia antenna is the formation of undesired hot spots in surrounding tissue [172]. This additional undesired heating often produces pain, burns, and blistering in the patient, which requires suspending or terminating the treatment. The patient does not receive general anesthetics during the hyperthermia treatment in order to provide direct verbal feedback of any pain. Thus, techniques for reducing undesired hot spots are necessary in hyperthermia treatment.

Special issues on acoustic and electromagnetic hyperthermia treatment of cancer have been published [41, 44, 46]. Electromagnetic transmitting antennas and arrays of antennas in the frequency band of about 60 to 2500 MHz have been investigated in localizing heating of malignant tumors within a target body. Many studies have been conducted to attempt to produce improved therapeutic field distributions with hyperthermia phased arrays [43-63]. Many of these studies would require invasive techniques to optimize the radiation pattern in tissue. Pretreatment planning is sometimes discussed in

Figure 1.8 An early conceptual drawing of a four-horn radiofrequency array applicator for deep-tumor heat treatment from a report by von Hippel, et al. [171, p. 17].

the literature in terms of controlling phased array hyperthermia sessions where patients are heated [49, 50]. This approach generally does not achieve desired heating characteristics since theoretical treatments and actual treatments can differ significantly. During hyperthermia treatments, phase drift in the phase shifters and power amplifiers as well as in the cables, and connectors and human body itself can lead to significant phase focusing errors [71]. An adaptive phased array with real-time feedback and control is a potentially viable approach for clinical treatments.

A number of studies by the author and collaborators [84-86, 90, 91] have investigated the theoretical benefit of using adaptive nulling with noninvasive auxiliary dipole sensors to reduce the field intensity at selected positions in the target body while maintaining a desired focus at a tumor. Measurements show that noninvasive electric-field probes can be used to eliminate hot spots that are shallow with respect to the phantom target tissue [85, 86]. It is important to note that elimination of deep tissue hot spots may require electric-field nulling with minimally invasive sensors. Array transmit weights can be adaptively controlled to maximize the tumor temperature (or RF power delivered to the tumor) while minimizing the surrounding tissue temperature (or RF power delivered to the surrounding tissue). This book addresses the potential benefits of using adaptive phased arrays with noninvasive or minimally invasive E-field probe sensors to reduce the field intensity at selected positions in the target body while maintaining a desired focus at a tumor.

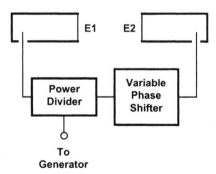

Figure 1.9 An early conceptual drawing of an opposing radiofrequency phased array applicator for heat treatment of cancer. Redrawn from von Hippel [171, p. 19]. In this diagram, E1 and E2 refer to the electric fields generated by the two opposing applicators.

1.5 ADAPTIVE PHASED ARRAY THERMOTHERAPY CONCEPT

In 1991, the author introduced an adaptive phased array approach to treating cancer [84, 173]. The concept of a noninvasive or minimally invasive adaptive-nulling hyperthermia system is shown in Figure 1.10. Theoretically, to generate the desired field distribution in a clinical adaptive phased array hyperthermia system, receiving sensors are positioned as close as possible to the focus (tumor site) and to where high temperatures are to be avoided (such as near the spinal cord, scar tissue, and skin). This book will show in deep hyperthermia simulated treatments in phantoms, for an annular array configuration the receiving sensors can be located noninvasively on the surface (skin) of the target. Initially, the adaptive phased array thermotherapy system is focused to produce the required field intensity at the tumor. Depending on the tumor site, a minimally invasive probe or intracavitary probe is required to achieve the optimum focus at depth. To avoid undesired hot spots, it is necessary to minimize the power received at the desired null positions and to constrain the array weights to deliver a required amount of transmitted or focal region power. The adaptive array weights (with gain g and phase ϕ) can be controlled by either the sample matrix inversion (SMI) algorithm or a gradient search (GS) algorithm to rapidly (within seconds) form the nulls before a significant amount of tissue heating takes place. With this adaptive technique, it should be possible to avoid unintentional hot spots and maintain a therapeutic thermal dose distribution within the tumor.

A specific example of an eight-element hyperthermia ring phased array with a target cross section through the prostate is shown in Figure 1.11(a). [4]. Sullivan analyzed an eight-element radiofrequency dipole ring phased array hyperthermia system with a water bolus surrounding a 35-cm by 25-cm elliptical phantom having a 2-cm fat layer. The phantom was assumed to contain homogeneous phantom tissue with dielectric constant 70 and electrical conductivity 0.68 S/m. Computer simulations using the method of moments (MoM) and finite-difference time-domain (FDTD) techniques (described in Chapter 3) for the electric-field amplitude at 110 MHz along the major axis of the phantom target are shown in Figure 1.12, and good agreement is observed in the focal region. Both the method of moments and finite-difference time-domain analyses predict large-amplitude electric fields away from the focal region moving toward the surface. (Further details of the method of moments simulations are given in Chapter 5.) These large-amplitude electric fields can give rise to undesired hot spots and pain in normal healthy tissues away from the deep-seated tumor.

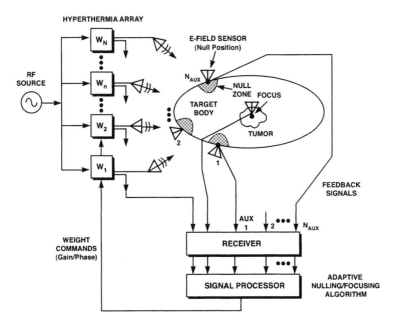

Figure 1.10 Noninvasive or minimally invasive adaptive phased array thermotherapy system concept. (From [86] with permission from Informa Healthcare, www.informaworld.com.)

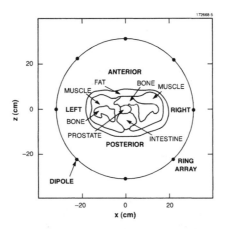

Figure 1.11 Transverse cross section through the prostate (redrawn from Sullivan 1990 [49]). An eight-element hyperthermia ring array of dipole elements is used to irradiate a target body with sufficient power to elevate the temperature of the target tissue to a therapeutic level.

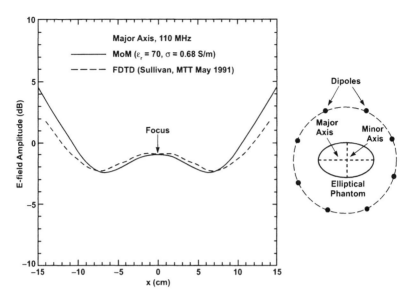

Figure 1.12 Computer simulations using the method of moments and finite-difference time-domain technique [50] for the electric field along the major axis of a phantom target.

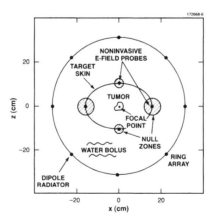

Figure 1.13 Elliptical phantom target used to model the body cross section. Shown are four noninvasive auxiliary E-field probes that are used in forming null zones with widths controlled by the individual null depths. The null zones extend into the body to reduce the internal electric field at specified locations.

Figure 1.13 shows an elliptical phantom target that is used to model the cross section of the body through the prostate. A noninvasive adaptive nulling system is achieved by placing auxiliary sensors $1, 2, \cdots, N_{aux}$ on the target skin as shown previously in Figure 1.10. The null zones centered at each auxiliary probe naturally extend into the elliptical target region to eliminate undesired hot spots. The width of each null zone is directly related to the depth of each null. The depth of each null (sometimes referred to as the amount of cancellation) is directly related to the signal-to-noise ratio at the sensor position. A low SNR produces a small amount of nulling, and a high SNR produces a large amount of nulling. The resolution or minimum spacing between the focus and null position is normally equal to the half-power beamwidth of the antenna. The resolution is enhanced somewhat by using weak nulls whenever the separation between the null and focus is closer than the half-power beamwidth in tissue.

Figure 1.14(a) shows a calculated two-dimensional thermal distribution for a 120-MHz RF hyperthermia ring array with eight adaptively controlled dipole elements transmitting uniformly, in amplitude and phase, through a constant-temperature water bolus into a homogeneous elliptical target region. The tumor site is assumed to be at the center of the ellipse. The thermal distribution in Figure 1.14(a) contains two hot spots to the left and right of the central focus. Note: a similar thermal distribution has been reported in the literature by Field and Hand [43, p. 292, 293]). In Figure 1.14(b), adaptive nulling at four independent surface positions (refer to Figure 1.13) is in effect and, due to the finite widths of the nulls, the two hot spots are eliminated. The calculated results in Figure 1.14 are based on a homogeneous moment method electric-field model as discussed in Chapters 3 and 5 and a thermal analysis model as based on the theory presented in Chapter 4.

Starting in 1990, the author investigated deep-heating adaptive phased array thermotherapy techniques for tumors located in the torso. Computer simulations and phantom measurements of adaptive phased array thermotherapy for deep tumors in the torso were performed. The author also investigated an adaptive phased array specialized to the case of focused microwave thermotherapy for treating breast cancer, which was evaluated by computer simulations, phantom measurements, animal studies, and human clinical trials [87-89, 93-98, 174]. An adaptive phased array focused microwave thermotherapy system used in clinical studies for treatment of cancer in the intact breast is shown in Figure 1.15. Preclinical testing in phantoms and in animals, and results for four clinical studies for this adaptive phased array thermotherapy system are presented in Chapters 8 and 9, respectively.

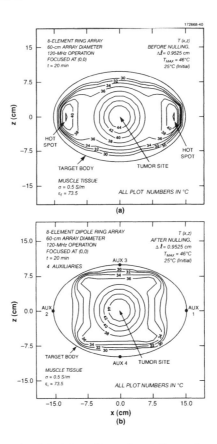

Figure 1.14 Computer simulated two-dimensional thermal patterns in an elliptical phantom target irradiated by an eight-channel ring phased array, as described in Chapter 5. The incident RF power distribution is at 120 MHz and the initial temperature of the phantom is 25°C. (a) Before adaptive nulling; hot spots on the left and right sides of the target are present. (b) After adaptive nulling; hot spots are eliminated. (From [86] with permission from Informa Healthcare, www.informaworld.com.)

1.6 SUMMARY

Adaptive phased array antennas that are used in radar and communications systems can be tested in the focused near-field region. A focused radiation beam maximum can be formed in the near-field of the phased array radar while forming near-field nulls, or pattern minima. Adaptive nulling in phased array radar systems is usually implemented in a receive mode in which external interference (jamming) is nulled so that the radar can detect desired targets. However, adaptive phased arrays can also be used in a transmit mode that has

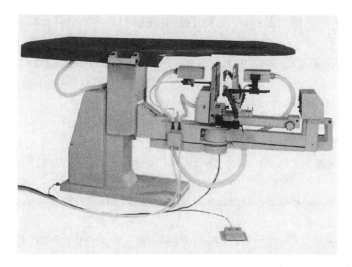

Figure 1.15 An adaptive phased array focused microwave thermotherapy system used in clinical studies for treatment of both early-stage invasive breast cancer and large breast carcinomas in the intact breast, as described in Chapters 8 and 9. (Photograph courtesy of Celsion (Canada) Limited.)

application in the hyperthermia treatment of cancer.

A background for cancer has been given, and it was observed that cancer burden is rising worldwide, and improved treatments are desired. Hyperthermia, or thermotherapy, is a method for treating cancer in which heat is used alone or in combination with radiation therapy and chemotherapy. Hyperthermia has been evaluated in randomized clinical trials and a statistically significant improvement in both tumor complete response and in patient survival has been demonstrated. Electromagnetic phased arrays operating at radio and microwave frequencies can be used to heat deep and semideep cancers, but it is difficult to heat tumors reliably without burning surrounding skin and body tissues. An adaptive transmit hyperthermia phased array antenna concept for treating cancer has been described in this chapter. The concept is intended as a primarily noninvasive treatment approach in which a deep tumor is heated with a focused electromagnetic field while nullifying the electromagnetic field on the surface of the body. Since the nulled electromagnetic field has a finite width, the null can penetrate into the body, providing some protection of internal body tissues. An adaptive phased array focused microwave thermotherapy system has been developed for treating breast cancer. In the next chapter, algorithms for implementing adaptive phased array thermotherapy are given.

References

[1] Schelkunoff, S.A., and H.T. Friis, *Antennas: Theory and Practice*, New York: Wiley, 1952.

[2] Kraus, J.D., *Antennas*, 2nd ed., New York: McGraw-Hill, 1988.

[3] Johnson, R.C., and H. Jasik, *Antenna Engineering Handbook*, 2nd ed., New York: McGraw-Hill, 1984.

[4] Mailloux, R.J., "Phased Array Theory and Technology," *Proc. IEEE*, Vol. 70, No. 3, 1982, pp. 246 - 291.

[5] Brookner, E., "Phased-Array Radars," *Scientific American*, Vol. 252, No. 2, 1985, pp. 94-102.

[6] Mailloux, R.J., *Phased Array Antenna Handbook*, Norwood, MA: Artech House, 1994.

[7] Hansen, R.C., *Phased Array Antennas*, New York: Wiley, 1998.

[8] Hansen, R.C., (ed.), *Significant Phased Array Antenna Papers*, Dedham, MA: Artech House, 1973.

[9] Oliner, A.A., and G.H. Knittel, (eds.), *Phased Array Antennas*, Dedham, MA: Artech House, 1972.

[10] Amitay, N., V. Galindo, and C.P. Wu, *Theory and Analysis of Phased Array Antennas*, New York: Wiley, 1972.

[11] Brookner, E., (ed.), *Practical Phased-Array Antenna Systems*, Norwood, MA: Artech House, 1991.

[12] Hansen, R.C., *Microwave Scanning Antennas, Vol. II: Array Theory and Practice*, Los Altos, CA: Peninsula Publishing, 1985.

[13] Bhattacharyya, A.K., *Phased Array Antennas*, New York: Wiley, 2006.

[14] Fenn, A.J., et al., "The Development of Phased Array Radar Technology," *Lincoln Laboratory Journal*, Vol. 12, No. 2, 2000, pp. 321-340.

[15] Allen, J.L., et al., "Phased Array Radar Studies, 1 July 1959 to 1 July 1960," Technical Report 228, Lincoln Laboratory, Lexington, MA: August 12, 1960, DTIC No. AD-0249470.

[16] Allen, J.L., et al., "Phased Array Radar Studies, 1 July 1, 1960 to 1 July 1961," Technical Report 236, Lincoln Laboratory, Lexington, MA: November 13, 1961, DTIC No. AD-271724.

[17] Allen, J.L., et al., "Phased Array Radar Studies, July 1, 1961 to July 1, 1963," Technical Report 299, Lincoln Laboratory, Lexington, MA: February 20, 1963, DTIC No. AD-417572.

[18] Allen, J.L., et al., "Phased Array Radar Studies, July 1, 1963 to July 1, 1964," Technical Report 381, Lincoln Laboratory, Lexington, MA: March 31, 1965, DTIC No. AD-629363.

[19] Allen, J.L., "Theory of Array Antennas (with Emphasis on Radar Applications)," Technical Report 323, Lincoln Laboratory, Lexington, MA: July 25, 1963, DTIC No. AD-422945.

[20] Fenn, A.J., *Adaptive Antennas and Phased Arrays for Radar and Communications,* Dedham, MA: Artech House, 2008.

[21] Reed, I.S., J.D. Mallet, and L.E. Brennan, "Rapid Convergence Rate in Adaptive Arrays," *IEEE Trans. Aerospace and Electronic Systems*, Vol. AES-10, No. 6, Nov. 1974, pp. 853-863.

[22] Howells, P.W., "Explorations in Fixed and Adaptive Resolution at GE and SURC," *IEEE Trans. Antennas Propagat.*, Vol. 24, No. 5, 1976, pp. 575-584.

[23] Applebaum, S.P., "Adaptive Arrays," *IEEE Trans. Antennas Propagat.*, Vol. 24, No. 5, 1976, pp. 585-598.

[24] Gabriel, W.F., "Adaptive Arrays – An Introduction," *Proc. IEEE,* Vol. 64, 1976, pp. 239-271.

[25] Monzingo, R.A., and T.W. Miller, *Introduction to Adaptive Arrays*, New York: Wiley, 1980.

[26] Compton, Jr., R.T., *Adaptive Antennas, Concepts and Performance*, Upper Saddle River, NJ: Prentice-Hall, 1988.

[27] Weiner, M.M., *Adaptive Antennas and Receivers*, Boca Raton, FL: CRC Press, 2006.

[28] Chandran, S., (ed.), *Adaptive Antenna Arrays: Trends and Applications*, Berlin: Springer-Verlag, 2004.

[29] Manolakis, D.G., V.K. Ingle, and S.M. Kogon, *Statistical and Adaptive Signal Processing: Spectral Estimation, Signal Modeling, Adaptive Filtering, and Array Processing,* Norwood, MA: Artech House, 2005.

[30] Nitzberg, R., *Adaptive Signal Processing for Radar*, Norwood, MA: Artech House, 1992.

[31] Farina, A., *Antenna-Based Signal Processing Techniques for Radar Systems,* Norwood, MA: Artech House, 1992.

[32] Hudson, J.E., *Adaptive Array Principles*, New York: Peter Peregrinus, 1981.

[33] Haykin, S.S., and A. Steinhardt, (eds.), *Adaptive Radar Detection and Estimation,* New York: Wiley, 1992.

[34] Allen, B., and M. Ghavami, *Adaptive Array Systems: Fundamentals and Applications,* New York: Wiley, 2005.

[35] Fenn, A.J., "Theory and Analysis of Near Field Adaptive Nulling," *1986 IEEE Antennas and Propagation Symposium Digest,* Vol. 2, New York: 1986, pp. 579-582.

[36] Fenn, A.J., "Theory and Analysis of Near Field Adaptive Nulling," *Proc Asilomar Conf Signals, Systems and Computers,* Computer Society Press of the IEEE, Washington, D.C.: Nov 10-12, 1986, pp. 105-109.

[37] Fenn, A.J., "Evaluation of Adaptive Phased Array Far-Field Nulling Performance in the Near-Field Region," *IEEE Trans. Antennas Propagat.*, Vol. 38, No. 2, 1990, pp. 173-185.

[38] Fenn, A.J., H.M. Aumann, F.G. Willwerth, and J.R. Johnson, "Focused Near-Field Adaptive Nulling: Experimental Investigation," *1990 IEEE Antennas Propagation Soc. Int. Symp. Digest,* Vol. 1, May 7-11, 1990, pp. 186-189.

[39] Fenn, A.J., "Analysis of Phase-Focused Near-Field Testing for Multiphase-Center Adaptive Radar Sysems," *IEEE Trans. Antennas Propagat.*, Vol. 40, No. 8, 1992, pp. 878-887.

[40] Fenn, A.J., "Moment Method Analysis of Near Field Adaptive Nulling," *IEE Sixth Int. Conf. on Antennas and Propagation, ICAP 89,* April 4-7, 1989, pp. 295-301.

[41] Cheung, A.Y., and G.M. Samaras, (eds.), "Special Issue on Hyperthermia Treatment of Cancer," *J Microwave Power*, Vol. 16, No. 2, 1981.

[42] Guy, A.W., "History of Biological Effects and Medical Applications of Microwave Energy," *IEEE Trans. on Microwave Theory and Techniques,* Vol. MTT-32, No. 9, 1984, pp. 1182-1200.

[43] Field, S.B. and J.W. Hand, (eds.), *An Introduction to the Practical Aspects of Clinical Hyperthermia,* London: Taylor & Francis, 1990.

[44] Strohbehn, J.W., T.C. Cetas, and G.M. Hahn, (eds.), "Special Issue on Hyperthermia and Cancer Therapy," *IEEE Trans. on Biomedical Engineering,* Vol. BME-31, No. 1, January 1984.

[45] Gauthrie, M., (ed.), *Methods of External Hyperthermic Heating,* New York, NY: Springer-Verlag, 1990.

[46] Lin, J.C., (ed.), "Special Issue on Phased Arrays for Hyperthermia Treatment of Cancer," *IEEE Trans. on Microwave Theory and Techniques,* Vol. MTT-34, No. 5, 1986.

[47] Sathiaseelan, V., M.F. Iskander, G.C.W. Howard, and N.M. Bleehen, "Theoretical Analysis and Clinical Demonstration of the Effect of Power Control Using the Annular Phased-Array Hyperthermia System," *IEEE Trans. on Microwave Theory and Techniques,* Vol. MTT-34, No. 5, 1986, pp. 514-519.

[48] Sathiaseelan, V., "Potential for Patient-Specific Optimization of Deep Heating Patterns Through Manipulation of Amplitude and Phase," *Strahlentherapie Onkologie*, Vol. 165, No. 10, 1989, pp. 743-745.

[49] Sullivan, D., "Three-Dimensional Computer Simulation in Deep Regional Hyperthermia Using the Finite-Difference Time-Domain Method," *IEEE Trans. Microwave Theory and Techniques,* Vol. MTT-38, No. 2, 1990, pp. 204-211.

[50] Sullivan, D., "Mathematical Methods for Treatment Planning in Deep Regional Hyperthermia," *IEEE Trans. Microwave Theory and Techniques*, Vol. 39, No. 5, 1991, pp. 864-872.

[51] Trembly, B.S., A.H. Wilson, M.J. Sullivan, A.D. Stein, T.Z. Wong, and J.W. Strohbehn, "Control of the SAR Pattern Within an Interstitial Microwave Array Through Variation of Antenna Driving Phase," *IEEE Trans. Microwave Theory and Techniques,* Vol. MTT-34, No. 5, 1986, pp. 568-571.

[52] Sato, G., C. Shibata, S. Sekimukai, H. Wakabayashi, K. Mitsuka, and K. Giga, "Phase-Controlled Circular Array Heating Equipment for Deep-Seated Tumors: Preliminary

Experiments," *IEEE Trans. Microwave Theory and Techniques,* Vol. MTT-34, No. 5, 1986, pp. 520-525.

[53] Cudd, P.A., A.P. Anderson, M.S. Hawley, and J. Conway, "Phased-Array Design Considerations for Deep Hyperthermia Through Layered Tissue," *IEEE Trans. Microwave Theory and Techniques,* Vol. MTT-34, No. 5, 1986, pp. 526-531.

[54] Morita, N., T. Hamasaki, and N. Kumagai, "An Optimal Excitation Method in Multiapplicator Systems for Forming a Hot Zone Inside the Human Body," *IEEE Trans. Microwave Theory and Techniques,* Vol. MTT-34, No. 5, 1986, pp. 532-538.

[55] De Wagter, C., "Optimization of Simulated Two-Dimensional Temperature Distributions Induced by Multiple Electromagnetic Applicators," *IEEE Trans. Microwave Theory and Techniques,* Vol. MTT-34, No. 5, 1986, pp. 589-596.

[56] Knudsen, M., and U. Hartmann, "Optimal Temperature Control with Phased-Array Hyperthermia System," *IEEE Trans. Microwave Theory and Techniques,* Vol. MTT-34, No. 5, 1986, pp. 597-603.

[57] Babbs, C.F., V.A. Vaguine, and J.T. Jones, "A Predictive-Adaptive, Multipoint Feedback Controller for Local Heat Therapy of Solid Tumors," *IEEE Trans. Microwave Theory and Techniques,* Vol. MTT-34, No. 5, 1986, pp. 604-611.

[58] Roemer, R.B., "Physical and Engineering Aspects of Hyperthermia," *Proc. 8th Int. Congress of Radiation Research,* Vol. 2, 1987, pp. 948-953.

[59] Roemer, R.B., K. Hynynen, C. Johnson, and R. Kress, "Feedback Control and Optimization of Hyperthermia Heating Patterns: Present Status and Future Needs," *IEEE Eighth Annual Conf. of the Engineering in Medicine and Biology Society,* 1986, pp. 1496-1499.

[60] Boag, A., and Y. Leviatan, "Optimal Excitation of Multiapplicator Systems for Deep Regional Hyperthermia," *IEEE Trans. Biomedical Engineering,* Vol. BME-37, No. 10, 1990, pp. 987-995.

[61] Loane III, J.T., and S.W. Lee, "Gain Optimization of a Near-Field Focusing Array for Hyperthermia Applications," *IEEE Trans. Microwave Theory and Techniques,* Vol. 37, No. 10, 1989, pp. 1629-1635.

[62] Anderson, A.P., and M. Melek, "Feasibility of Focused Microwave Array System for Tumour Irradiation," *Electronics Letters,* Vol. 15, No. 18, August 30, 1979, pp. 564-565.

[63] Turner, P.F., A. Tumeh, and T. Schaefermeyer, "BSD-2000 Approach for Deep Local and Regional Hyperthermia: Physics and Technology," *Strahlentherapie Onkologie,* Vol. 165, No. 10, 1989, pp. 738-741.

[64] Strohbehn, J.W., and R.B. Roemer, "A Survey of Computer Simulations of Hyperthermia Treatments," *IEEE Trans. on Biomedical Engineering,* Vol. BME-31, No. 1, 1984, pp. 136-149.

[65] Ocheltree, K.B., and L.A. Frizzell, "Determination of Power Deposition Patterns for Localized Hyperthermia: A Transient Analysis," *Int. J. Hyperthermia,* Vol. 4, No. 3, 1988, pp. 281-296.

[66] Shimm, D.S., T.C. Cetas, J.R. Oleson, E.R. Gross, D.N. Buechler, A.M. Fletcher, and S.E.

Dean, "Regional Hyperthermia for Deep-Seated Malignancies Using the BSD Annular Array," *Int. J. Hyperthermia*, Vol. 4, No. 2, 1988, pp. 159-170.

[67] Zhang, Y., W.T. Joines, and J.R. Oleson, "The Calculated and Measured Temperature Distribution of a Phased Interstitial Antenna Array," *IEEE Trans. on Microwave Theory and Techniques*, Vol. 38, No. 1, 1990, pp. 69-77.

[68] Myerson, R.J., L. Leybovich, B. Emami, P.W. Grigsby, W. Straube, and D. von Gerichten, "Phantom Studies and Preliminary Clinical Experience with the BSD 2000," *Int. J. Hyperthermia*, Vol. 7, No. 6, 1991, pp. 937-951.

[69] Wust, P., J. Nadobny, R. Felix, P. Deulhard, A. Louis, and W. John, "Strategies for Optimized Application of Annular-Phased-Array Systems in Clinical Hyperthermia," *Int. J. Hyperthermia*, Vol. 7, No. 1, 1991, pp. 157-173.

[70] Nikita, K.S., N. Maratos, and N.K. Uzunoglu, "Optimum Excitation of Phases and Amplitudes in a Phased Array Hyperthermia System," *Int. J. Hyperthermia*, Vol. 8, No. 4, 1992, pp. 515-528.

[71] Straube, W.L., E.G. Moros, and R.J. Myerson, "Phase Stability of a Clinical Phased Array System for Deep Regional Hyperthermia," *Int. J. Hyperthermia*, Vol. 11, No. 1, 1995, pp. 87-93.

[72] Paulsen, K.D., S. Geimer, J. Tang, and W.E. Boyse, "Optimization of Pelvic Heating Rate Distributions with Electromagnetic Phased Arrays," *Int. J. Hyperthermia*, Vol. 15, No. 3, 1999, pp. 157-186.

[73] Seebass, M., R. Beck, J. Gellermann, J. Nadobny, and P. Wust, "Electromagnetic Phased Arrays for Regional Hyperthermia: Optimal Frequency and Antenna Arrangement," *Int. J. Hyperthermia*, Vol. 17, No. 4, 2001, pp. 321-336.

[74] Das, S.K., E.A. Jones, and T.V. Samulski, "A Method of MRI-Based Thermal Modelling for a RF Phased Array," *Int. J. Hyperthermia*, Vol. 17, No. 6, 2001, pp. 465-482.

[75] Wiersma, J., R.A.M. van Maarseveen, and J.D.P. van Dijk, "A Flexible Optimization Tool for Hyperthermia Treatments with RF Phased Array Systems," *Int. J. Hyperthermia*, Vol. 18, No. 2, 2002, pp. 73-85.

[76] Shi, G., and W.T. Joines, "Design and Analysis of Annular Antenna Arrays with Different Reflectors," *Int. J. Hyperthermia*, Vol. 20, No. 6, 2004, pp. 625-636.

[77] Kongsli, J., B.T. Hjertaker, and T. Frøystein, "Evaluation of Power and Phase Accuracy of the BSD Dodek Amplifier for Regional Hyperthermia Using an External Vector Voltmeter Measurement System," *Int. J. Hyperthermia*, Vol. 22, No. 8, 2006, pp. 657-671.

[78] Paulides, M.M., J.F. Bakker, A.P.M. Zwamborn, and G.C. van Rhoon, "A Head and Neck Hyperthermia Applicator: Theoretical Antenna Array Design," *Int. J. Hyperthermia*, Vol. 23, No. 1, 2007, pp. 59-67.

[79] Jones, E., A.A. Secord, L.R. Prosnitz, T.V. Samulski, J.R. Oleson, A. Berchuck, D. Clarke-Pearson, J. Soper, M. W. Dewhirst, and Z. Vujaskovic, "Intra-Peritoneal Cisplatin and Whole Abdomen Hyperthermia for Relapsed Ovarian Carcinoma," *Int. J. Hyperthermia*, Vol. 22, No. 2, 2006, pp. 161-172.

[80] Fatehi, D., J. van der Zee, M. de Bruijne, M. Franckena, and G.C. van Rhoon, "RF-power and Temperature Data Analysis of 444 Patients with Primary Cervical Cancer: Deep Hyperthermia Using the Sigma-60 Applicator is Reproducible," *Int. J. Hyperthermia*, Vol. 23, No. 8, 2007, pp. 623-643.

[81] Fatehi, D., and G.C. van Rhoon "SAR Characteristics of the Sigma-60-Ellipse Applicator," *Int. J. Hyperthermia*, published online January 9, 2008, (www.thermalmedicine.org).

[82] Fenn, A.J., "Adaptive Nulling Hyperthermia Array," US Patent No. 5,251,645, October 12, 1993.

[83] Fenn, A.J., "Adaptive Focusing and Nulling Hyperthermia Annular and Monopole Phased Array Applicators," US Patent No. 5,441,532, August 15, 1995.

[84] Fenn, A.J., "Adaptive Hyperthermia for Improved Thermal Dose Distribution," In: *Radiation Research: A Twentieth Century Perspective,* Vol. 1 (Congress Abstracts), Chapman J.D., W.C. Dewey, G.F. Whitmore, (eds.), San Diego, Calif.: Academic Press, 1991, p. 290.

[85] Fenn, A.J., and G.A. King, "Experimental Investigation of an Adaptive Feedback Algorithm for Hot Spot Reduction in Radio-Frequency Phased-Array Hyperthermia," *IEEE Trans Biomed Eng.*, Vol. 43, No. 3, 1994, pp. 273-280.

[86] Fenn, A.J., and G.A. King, "Adaptive Radio Frequency Hyperthermia Phased Array System for Improved Cancer Therapy: Phantom Target Measurements," *Int J Hyperthermia*, Vol. 10, No. 2, 1994, pp. 189-208.

[87] Fenn, A.J., B.A. Bornstein, G.K. Svensson, and H.F. Bowman, "Minimally Invasive Monopole Phased Arrays for Hyperthermia Treatment of Breast Carcinomas: Design and Phantom Tests," *Int. Symp. on Electromagnetic Compatibility*, Sendai, Japan, Vol. 10, No. 2, 1994, pp. 566-569.

[88] Fenn, A.J., "Minimally Invasive Monopole Phased Arrays for Hyperthermia Treatment of Breast Cancer," In: *Proc. 1994 Int. Symp. on Antennas,* Nice, France: November 8-10, 1994, pp. 418-421.

[89] Fenn, A.J., "Minimally Invasive Monopole Phased Array Hyperthermia Applicators and Method for Treating Breast Carcinomas," US Patent No. 5,540,737, July 30, 1996.

[90] Sathiaseelan, V., A.J. Fenn, and A. Taflove, "Recent Advances in External Electromagnetic Hyperthermia," In: Chapter 10 of *Advances in Radiation Treatment,* Mittal, B.B., J.A. Purdy, and K.K. Ang, (eds.), Boston, Massachusetts: Kluwer Academic Publishers, 1998, pp. 213-245.

[91] Fenn, A.J., V. Sathiaseelan, G.A. King, and P.R. Stauffer, "Improved Localization of Energy Deposition in Adaptive Phased Array Hyperthermia Treatment of Cancer," *J Oncol Management*, Vol. 7, No. 2, 1998, pp. 22-29.

[92] Fenn, A.J., "Thermodynamic Adaptive Phased Array System for Activating Thermosensitive Liposomes in Targeted Drug Delivery," US Patent No. 5,810,888, September 22, 1998.

[93] Fenn, A.J., G.L. Wolf, and R.M. Fogle, "An Adaptive Phased Array for Targeted Heating of Deep Tumors in Intact Breast: Animal Study Results," *Int J Hyperthermia*, Vol. 15,

No. 1, 1999, pp. 45-61.

[94] Gavrilov, L.R., J.W. Hand, J.W. Hopewell, and A.J. Fenn, "Pre-clinical Evaluation of a Two-Channel Microwave Hyperthermia System with Adaptive Phase Control in a Large Animal," *Int J Hyperthermia*, Vol. 15, No. 6, 1999, pp. 495-507.

[95] Gardner, R.A., H.I. Vargas, J.B. Block, C.L. Vogel, A.J. Fenn, G.V. Kuehl, and M. Doval, "Focused Microwave Phased Array Thermotherapy for Primary Breast Cancer," *Ann Surg Oncol*, Vol. 9, No. 4, 2002, pp. 326-332.

[96] Vargas, H.I., W.C. Dooley, R.A. Gardner, K.D. Gonzalez, S.H. Heywang-Kobrunner, and A.J. Fenn, "Focused Microwave Phased Array Thermotherapy for Ablation of Early-Stage Breast Cancer: Results of Thermal Dose Escalation," *Ann Surg Oncol*, Vol. 11, No. 2, 2004, pp. 139-146.

[97] Fenn, A.J., *Breast Cancer Treatment by Focused Microwave Thermotherapy*, Sudbury, MA: Jones and Bartlett, 2007.

[98] Vargas, H.I., W.C. Dooley, A.J. Fenn, M.B. Tomaselli, and J.K. Harness, "Study of Preoperative Focused Microwave Phased Array Thermotherapy in Combination With Neoadjuvant Anthracycline-Based Chemotherapy for Large Breast Carcinomas," *Cancer Therapy*, Vol. 5, 2007, pp. 401-408, published online (www.cancer-therapy.org), November 25, 2007.

[99] Mayhan, J.T., "Some Techniques for Evaluating the Bandwidth Characteristics of Adaptive Nulling Systems," *IEEE Trans. Antennas Propagat.*, Vol. 27, No. 3, 1979, pp. 363-373.

[100] Mayhan, J.T., A.J. Simmons, and W.C. Cummings, "Wide-Band Adaptive Antenna Nulling Using Tapped Delay Lines," *IEEE Trans. Antennas Propagat.*, Vol. 29, No. 6, 1981, pp. 923-936.

[101] Mayhan, J.T., "Adaptive Nulling with Multiple-Beam Antennas," *IEEE Trans. Antennas Propagat.*, Vol. 26, No. 2, 1978, pp. 267-273.

[102] Fenn, A.J., "Maximizing Jammer Effectiveness for Evaluating the Performance of Adaptive Nulling Array Antennas," *IEEE Trans. Antennas Propagat.*, Vol. 33, No. 10, 1985, pp. 1131-1142.

[103] Fenn, A.J., "Interference Sources and Degrees of Freedom in Adaptive Nulling Antennas," Technical Report 604, Lincoln Laboratory, Massachusetts Institute of Technology, May 12, 1982.

[104] Fenn, A.J., "Consumption of Degrees of Freedom in Adaptive Nulling Array Antennas," Technical Report 609, Lincoln Laboratory, Massachusetts Institute of Technology, October 12, 1982.

[105] Skolnik, M.I., *Introduction to Radar Systems*, 3rd ed., New York: McGraw-Hill, 2001.

[106] Skolnik, M.I., (ed.), *Radar Handbook*, 2nd ed., New York: McGraw-Hill, 1990, pp. 12.1-13.27.

[107] Scharfman, W.E., and G. August, "Pattern Measurements of Phased-Arrayed Antennas by Focusing into the Near Zone," In: *Phased Array Antennas (Proc. of the 1970 Phased*

Array Antenna Symposium), Oliner, A.A., and G.H. Knittel, (eds.), Dedham, MA: Artech House, 1972, pp. 344-350.

[108] Duck, F.A., *Physical Properties of Tissue: A Comprehensive Reference Book,* San Diego, CA: Academic Press, 1990.

[109] American Cancer Society, Cancer Facts & Figures 2007, Atlanta: American Cancer Society, 2007.

[110] World Health Statistics 2007, *World Health Organization*, Geneva, Switzerland, 2007.

[111] Mathers, C.D., and D. Loncar, "Projections of Global Mortality and Burden of Disease from 2002 to 2030," *PLoS Med*, Vol. 3, No. 11, 2006, pp. 2011-2030.

[112] Bray, F., B. Moller, "Predicting the Future Burden of Cancer," *Nature Reviews Cancer*, Vol. 6, 2006, pp. 63-74.

[113] Kantoff, P.W., (ed.), *Multidisciplinary Treatment for Prostate Cancer*, New York: CMP Medica, 2007.

[114] Chu, E., (ed.), *A Multidisciplinary Approach to the Treatment of Early Colorectal Cancer*, New York: CMP Medica, 2007.

[115] Marshall, J.L., (ed.), *New Treatment Paradigms in Colorectal Cancer*, New York: CMP Medica, 2006.

[116] Sandler, A.B., (ed.), *New Treatment Paradigms in Non-Small-Cell Lung Cancer*, New York: CMP Medica, 2006.

[117] Govindan, R., (ed.), *Recent Advances in Locally Advanced Non-Small-Cell Lung Cancer*, New York: CMP Medica, 2006.

[118] Posner, M.R., (ed.), *Options in the Treatment of Head and Neck Cancer*, New York: CMP Medica, 2006.

[119] Macdonald, J.S., (ed.), *Advances in the Management of Gastric Cancer*, New York: CMP Medica, 2006.

[120] Bukowski, R.M., (ed.), *New Treatment Paradigms in Renal Cell Carcinoma*, New York: CMP Medica, 2007.

[121] Engstrom, P.F., et al., "The NCCN Colon Cancer Clinical Practice Guidelines in Oncology," *J Nat Comprehensive Cancer Network*, Vol. 5, No. 9, 2007, pp. 884-925.

[122] Engstrom, P.F., et al., "The NCCN Rectal Cancer Clinical Practice Guidelines in Oncology," *J Nat Comprehensive Cancer Network*, Vol. 5, No. 9, 2007, pp. 940-981.

[123] Tempero, M., et al., "The NCCN Pancreatic Adenocarcinoma Cancer Clinical Practice Guidelines in Oncology," *J Nat Comprehensive Cancer Network*, Vol. 5, No. 10, 2007, pp. 998-1033.

[124] Morgan, R.J., et al., "The NCCN Ovarian Cancer Clinical Practice Guidelines in Oncology," *J Nat Comprehensive Cancer Network*, Vol. 4, No. 9, 2006, pp. 912-939.

[125] Carlson, R.W., et al., "The NCCN Invasive Breast Cancer Clinical Practice Guidelines in Oncology," *J Nat Comprehensive Cancer Network*, Vol. 5, No. 3, 2007, pp. 246-312.

[126] Mohler, J., "The NCCN Prostate Cancer Clinical Practice Guidelines in Oncology," *J Nat Comprehensive Cancer Network*, Vol. 5, No. 7, 2007, pp. 650-683.

[127] Falk, M.H., and R.D. Issels, "Hyperthermia and Oncology," *Int J Hyperthermia*, Vol. 17, No. 1, 2001, pp. 1-18.

[128] Hall, E.J., *Radiobiology for the Radiologist,* Philadelphia, PA: J.B. Lippincott, 1994, pp. 262-263.

[129] Perez, C.A., and L.W. Brady, *Principles and Practice of Radiation Oncology,* 2nd ed. Philadelphia, PA: J.B. Lippincott, 1992, pp. 396-397.

[130] Streffer, C., "Hyperthermia and the Therapy of Malignant Tumors," In: *Cancer Therapy by Hyperthermia and Radiation,* Streffer C., (ed.), New York: Springer-Verlag, 1987.

[131] Bicher, H.I., and D.F. Bruley, *Hyperthermia. Proceedings of the First Annual Meeting of the North American Hyperthermia Group,* August 23-25, 1981; Detroit, Michigan. New York, NY: Plenum Press, 1982.

[132] Hahn, G.M., *Hyperthermia and Cancer*, New York, NY: Plenum Press, 1982.

[133] Steeves, R.A., and B.R. Paliwal, (eds.), *Syllabus: A Categorical Course in Radiation Therapy: Hyperthermia,* 73rd Scientific Assembly and Annual Meeting of the Radiological Society of North America; November 29-December 4, 1987, Oak Brook, Illinois.

[134] Hill, R.P., and J.W. Hunt, *Hyperthermia in the Basic Science of Oncology,* Tannock, I.F. and R.P. Hill, (eds.), New York, NY: Pergamon Press, 1987, pp. 337-357.

[135] Hinklebein, W., G. Gruggmoser, R. Engelhardt, and M. Wannenmacher, (eds.), *Preclinical Hyperthermia, Recent Results in Cancer Research,* Vol. 109, New York, NY: Springer-Verlag, 1988.

[136] Matsuda, T., (ed.), *Cancer Treatment by Hyperthermia, Radiation and Drugs,* London: Taylor & Francis, 1993.

[137] Steeves, R.A., *Hyperthermia, The Radiologic Clinics of North America*, Philadelphia, Pa: WB Saunders Company, Vol. 27, No. 3, 1989.

[138] Seegenschmiedt, M.H., P. Fessenden, and C.C. Vernon, (eds.), *Thermoradiotherapy and Thermochemotherapy, Vol. 1, Biology, Physiology, and Physics,* Berlin: Springer, 1995.

[139] Seegenschmiedt, M.H., P. Fessenden, and C.C. Vernon, (eds.), *Thermoradiotherapy and Thermochemotherapy, Vol. 2, Clinical Applications,* Berlin: Springer, 1995.

[140] Lehmann, J.F., (ed.), *Therapeutic Heat and Cold,* 3rd ed. Baltimore, MD: Williams & Wilkins, 1982.

[141] Coughlin, C.T., Clinical Hyperthermic Practice: Interstitial Heating. In: *An Introduction to the Practical Aspects of Clinical Hyperthermia,* Field, S.B., and J.W. Hand, (eds.), London: Taylor & Francis; 1990, pp. 172-183.

[142] Kapp, D.S., and J.L. Meyer, Clinical Hyperthermic Practice: Non-Invasive Heating, In: *An Introduction to the Practical Aspects of Clinical Hyperthermia,* Field, S.B., J.W. Hand, (eds.), London: Taylor & Francis, 1990, pp. 143-171.

[143] Overgaard, J., "The Effect of Local Hyperthermia Alone, and in Combination with Radiation on Solid Tumors," In: *Cancer Therapy by Hyperthermia and Radiation, Proceedings of the 2nd International Symposium*, C. Streffer (ed.), June 2-4, 1977, Baltimore: Urban & Schwarzenberg, Inc., 1978, pp. 49-61.

[144] Overgaard, J., "Clinical Hyperthermia, an Update," *Proceedings of the 8th International Congress of Radiation Research*, Vol. 2, 1987, pp. 942-947.

[145] Lepock, J.R., "How Do Cells Respond to Their Thermal Environment?" *Int J Hyperthermia*, Vol. 21, No. 8, 2005, pp. 681-687.

[146] Dewhirst, M.W., B.L. Viglianti, M. Lora-Michiels, M. Hanson, P.J. Hoopes, "Basic Principles of Thermal Dosimetry and Thermal Thresholds for Tissue Damage from Hyperthermia," *Int J Hyperthermia*, Vol. 19, No. 3, 2003, pp. 267-294.

[147] Sapareto, S.A., and W.C. Dewey, "Thermal Dose Determination in Cancer Therapy," *Int J Rad Oncol Biol Phys*, Vol. 10, 1984, pp. 787-800.

[148] Roti Roti, J.L., "Cellular Responses to Hyperthermia (40-46°C): Cell Killing and Molecular Events," *Int J Hyperthermia*, Vol. 24, No. 1, 2008, pp. 3-15.

[149] Vernon, C.C., J.W. Hand, S.B. Field, et al., "Radiotherapy with or without Hyperthermia in the Treatment of Superficial Localized Breast Cancer: Results From Five Randomized Controlled Trials," *Int J Radiat Oncol Biol Phys*, Vol. 35, 1996, pp. 731-744.

[150] Valdagni, R., and M. Amichetti, "Report of Long Term Follow Up in a Randomized Trial Comparing Radiation Therapy and Radiation Therapy Plus Hyperthermia to Metastatic Lymph Nodes in Stage IV Head and Neck Patients," *Int J Radiat Oncol Biol Phys*, Vol. 28, 1993, pp. 163-169.

[151] Overgaard, J., D. Gonzalez Gonzalez, M.C.C.H. Hulshof, et al., "Hyperthermia as an Adjuvant to Radiation Therapy of Recurrent or Metastatic Malignant Melanoma. A Multicentre Randomized Trial by the European Society for Hyperthermic Oncology," *Int. J Hyperthermia*, Vol. 12, No. 1, 1996, pp. 3-20.

[152] van der Zee, J., D. Gonzalez Gonzalez, G.C. van Rhoon, et al., "Comparison of Radiotherapy Alone with Radiotherapy Plus Hyperthermia in Locally Advanced Pelvic Tumors: A Prospective, Randomised, Multicentre Trial," *Lancet*, Vol. 355, 2000, pp. 1119-1125.

[153] Sneed, P.K., P.R. Stauffer, M.W. McDermott, C.J. Diederich, K.R. Lamborn, et al., "Survival Benefit of Hyperthermia in a Prospective Randomized Trial of Brachytherapy Boost +/- Hyperthermia for Glioblastoma Multiforme," *Int J Radiat Oncol Biol Phys*, Vol. 40, No. 2, 1998, pp. 287-295.

[154] Jones, E.L., J.R. Oleson, L.R. Prosnitz, T.V. Samulski, Z. Vujaskovic, D. Yu, L.L. Sanders, and M.W. Dewhirst, "Randomized Trial of Hyperthermia and Radiation for Superficial Tumors," *J Clin Oncol*, Vol. 23, 2005, pp. 3079-3085.

[155] Harima, Y., K. Nagata, K. Harima, et al., A Randomized Clinical Trial of Radiation Therapy Versus Thermoradiotherapy in Stage IIIB Cervical Carcinoma, *Int J Hyperthermia*, Vol. 17, No. 2, 2001, pp. 97-105.

[156] Sugimachi, K., H. Kuwano, H. Ide, T. Toge, M. Saku, and Y. Oshiumi, "Chemotherapy Combined with or without Hyperthermia for Patients with Oesophageal Carcinoma: A

Prospective Randomized Trial," *Int J Hyperthermia*, Vol. 10, No. 4, 1994, pp. 485-493.

[157] Stauffer, P.R., "Evolving Technology for Thermal Therapy of Cancer," *Int. J. Hyperthermia,* Vol. 21, No. 8, 2005, pp. 731-744.

[158] Diederich, C.J., "Thermal Ablation and High-Temperature Thermal Therapy: Overview of Technology and Clinical Implementation," *Int J Hyperthermia,* Vol. 21, No. 8, 2005, pp. 745-753.

[159] Barqawi, A.B., and E.D. Crawford, "Emerging Role of HIFU as a Noninvasive Ablation Method to Treat Localized Prostate Cancer," *Oncology,* Vol. 22, No. 2, 2008, pp. 123-129.

[160] Ellis, L.M., S.A. Curley, and K.K. Tanabe, (eds.) *Radiofrequency Ablation for Cancer: Current Indications, Techniques, and Outcomes,* New York: Springer-Verlag, 2004.

[161] Agnese, D.M., and W.E. Burak, "Ablative Approaches to the Minimally Invasive Treatment of Breast Cancer," *Cancer J,* Vol. 11, 2005, pp. 77-82.

[162] Huston, T.L., and R.M. Simmons, "Ablative Therapies for the Treatment of Malignant Diseases of the Breast," *Am J Surg,* Vol. 189, 2005, pp. 694-701.

[163] Vitrogan, D., *Elements of Electric and Magnetic Circuits,* San Francisco: Rinehart Press, 1971, pp. 31-34.

[164] Gilman, A.G., L.S. Goodman, T.W. Rall, and F. Murad, (eds.), *The Phamacological Basis of Therapeutics,* 7th ed., New York: MacMillan, 1985, pp. 323-328.

[165] Zaffaroni, N., G. Fiorentini, and U. De Giorgi, "Hyperthermia and Hypoxia: New Developments in Anticancer Therapy," *European J Surg Oncol*, 2001, pp. 340-342.

[166] Vujaskovic, Z., E.L. Rosen, K.L. Blackwell, E.L. Jones, D.M. Brizel, L.R. Prosnitz, T.V. Samulski, and M.W. Dewhirst, "Ultrasound Guided pO2 Measurement of Breast Cancer Reoxygenation after Neoadjuvant Chemotherapy and Hyperthermia Treatment," *Int J Hyperthermia,* Vol. 19, No. 5, 2003, pp. 498-506.

[167] Jain, R.K., "Barriers to Drug Delivery in Solid Tumors," *Scientific American*, Vol. 276, July 1994, pp. 58-65.

[168] Song, C.W., H.J. Park, C.K. Lee, and R. Griffin, "Implications of Increase Tumor Blood Flow and Oxygenation Caused by Mild Temperature Hyperthermia in Tumor Treatment," *Int J. Hyperthermia,* Vol. 21, No. 8, 2005, pp. 761-767.

[169] Dewhirst, M.W., Z. Vujaskovic, E. Jones, and D. Thrall, "Re-setting the Biologic Rationale for Thermal Therapy," *Int J. Hyperthermia*, Vol. 21, No. 8, 2005, pp. 779-790.

[170] Straube, W.L., E.E. Klein, E.G. Moros, D.A. Low, and R.J. Myerson, "Dosimetry and Techniques for Simultaneous Hyperthermia and External Beam Radiation Therapy," Vol. 17, No. 1, 2001, pp. 48-62.

[171] von Hippel, A.R., A.H. Runck, and W.B. Westphal, *Dielectric Analysis of Biomaterials*, Cambridge, MA: Laboratory for Insulation Research, Massachusetts Institute of Technology, Technical Report 13, AD-769 843, 1973.

[172] Wiersma, J., N. van Wieringen, H. Crezee, and J.D. van Dijk, "Delineation of Potential Hot Spots for Hyperthermia Treatment Planning Optimisation," *Int J Hyperthermia*, Vol. 23, No. 3, 2007, pp. 287-301.

[173] Fenn, A.J., "Non-Invasive Adaptive Nulling for Improved Hyperthermia Thermal Dose Distribution," *IEEE Engineering in Medicine and Biology Society Int Conf*, October 31 - November 3, 1991, Vol. 13, No. 2, 1991, pp. 976-977.

[174] Dooley, W.C., H.I. Vargas, A.J. Fenn, M.B. Tomaselli, and J.K. Harness, "Randomized Study of Preoperative Focused Microwave Phased Array Thermotherapy for Early-Stage Invasive Breast Cancer," *Cancer Therapy*, Vol. 6, 2008, published online (www.cancer-therapy.org), August 25, 2008, pp. 395-408.

2

Adaptive Phased Array Algorithms for Thermotherapy

2.1 INTRODUCTION

The concept of an adaptive phased array thermotherapy system [1-14] for heat treatment of cancer was described in Chapter 1. Referring to Figure 1.10, an adaptive phased array hyperthermia system relies on feedback signals and a computerized algorithm that can determine the desired amplitude and phase settings for each antenna element of the array such that a desired radiation pattern is achieved. The desired radiation pattern consists of a beam focused on a tumor with nulled or reduced radiation directed at surrounding healthy tissues. Two example adaptive array algorithms that apply for adaptive radar and communications [15-36], as well as hyperthermia treatment of cancer, are the sample matrix inversion (SMI) algorithm [1, 16-18, 24, 31, 36] and the gradient-search (GS) algorithm [24-38]. In this chapter, the sample matrix inversion and gradient-search algorithms are described in the context of an adaptive transmitting phased array antenna applied to treating cancer.

2.2 SAMPLE MATRIX INVERSION ALGORITHM

With the SMI algorithm applied to hyperthermia treatment of cancer [1], the fundamental quantities required to fully characterize the incident field for adaptive nulling purposes are the adaptive channel cross correlations. To implement this algorithm it is necessary to quantify the complex received voltage at a set of auxiliary probes. For example, the moment-method formulation (described in Chapter 3) allows computation of the complex-received voltage at a set of E-field sensor probes. Figure 2.1 shows a

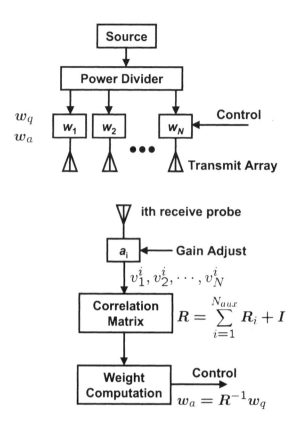

Figure 2.1 Block diagram of the sample matrix inversion (SMI) algorithm for an adaptive phased array hyperthermia system.

block diagram of the SMI algorithm that is used in the hyperthermia analysis presented in Chapter 5. Four quantitative performance measures are used in the computer simulations presented in Chapter 5: electric-field distribution, covariance matrix eigenvalues, adaptive transmit weights, and power cancellation. The calculation of these performance measures is described below.

Referencing Figures 1.10 and 2.1, let a spherical wavefront be incident at the ith probe antenna, due to each array element (radiating one at a time with a unity-amplitude reference signal), which results in a set of probe-received complex voltages denoted $v_1^i, v_2^i, \cdots, v_N^i$. The cross correlation R_{mn}^i of the received voltages due to the mth and nth adaptive transmit channels at the ith probe is given by [36, pp. 79-82]

$$R_{mn}^i = E(v_m v_n^*) \tag{2.1}$$

where $*$ means complex conjugate and $E(\cdot)$ means mathematical expectation (mean value). Because v_m and v_n represent voltages of the same waveform but at different times, R^i_{mn} is also referred to as an autocorrelation function. It should be noted that, for convenience, in (2.1) the superscript i in v_m and in v_n has been omitted.

Equation (2.1) can be expressed as the frequency average

$$R^i_{mn} = \frac{1}{B} \int_{f_1}^{f_2} v_m(f) v_n^*(f) \, df \tag{2.2}$$

where $B = f_2 - f_1$ is the nulling bandwidth and f is the frequency. It should be noted that $v_m(f)$ takes into account the transmit wavefront shape, which typically is approximately spherical for the hyperthermia application. For the special case of a continuous wave (CW) transmit waveform, as is normally used in hyperthermia, the cross correlation reduces to

$$R^i_{mn} = v_m(f_o) v_n^*(f_o) \tag{2.3}$$

where f_o is the transmit frequency of the hyperthermia array.

Let the channel or interference covariance matrix be denoted R. (Note: in hyperthermia, interference is used here to refer to the signals received at the auxiliary nulling probes. The undesired radiation, which can produce hot spots, can be thought of as interfering with the therapy.) If there are N_{aux} independent desired null positions or auxiliary probes, then the N_{aux}-channel covariance matrix is the sum of the covariance matrices observed at the individual probes; that is,

$$R = \sum_{i=1}^{N_{aux}} R_i + I \tag{2.4}$$

where R_i is the sample covariance matrix observed at the ith probe and I is the identity matrix used to represent the thermal noise level of the receiver.

Prior to generating an adaptive null, the channel weight vector, w, is chosen to synthesize a desired quiescent radiation pattern. In hyperthermia treatments, the desired radiation pattern would be focused on the tumor. When adaptive nulling is desired to remove hot spots, the optimum set of transmit weights to form an adaptive null (or nulls), denoted w_a and referred to as the adaptive weight vector, is computed by [15]

$$w_a = R^{-1} w_q \tag{2.5}$$

where $^{-1}$ means matrix inverse and w_q is the quiescent weight vector (or steering vector that points the array radiation pattern maximum in a desired

direction). During array calibration for focusing the array, the normalized
quiescent transmit weight vector, with element 1 radiating, is chosen to be
$w_q = (1, 0, 0, \cdots, 0)^T$, where ^{T} means transpose; that is, the transmit channel
weight of element 1 is unity and the remaining transmit channel weights
are zero. Similar weight settings are used to phase calibrate the remaining
transmit elements. Phase calibration can be accomplished in practice by
applying the complex conjugate operation to the array weights to remove the
element-to-element phase variation observed at the focal point. For a fully
adaptive annular array focused at the origin in ideal homogeneous tissue, the
normalized quiescent weight vector is simply a uniform weight vector, that
is, $w_q = (1, 1, 1, \cdots, 1)^T$. In clinical practice, for heterogeneous tissues; the
weight vector for focused radiation would be expected to deviate significantly
from this uniform quiescent weight vector. The weight vector is constrained to
deliver a required amount of power to the hyperthermia array or to the tumor.
For simplicity, the normalized weights can be constrained such that

$$w^\dagger w = \sum_{n=1}^{N} |w_n|^2 = 1 \tag{2.6}$$

where w_n is the transmit weight for the nth element and † means complex
conjugate transpose.

Let $y = w^\dagger v$ be the output signal from the array of probes. Then the
summation of power received at the probes is given by

$$p = |y|^2 = yy^* = w^\dagger vv^\dagger w = w^\dagger Rw \tag{2.7}$$

where w is a complex weight vector. The interference-plus-noise-to-noise
ratio, denoted INR, is computed as the ratio of the auxiliary probe array output
power [defined in (2.7)] with the transmit signal present to the probe array
output power with only receiver noise present; that is,

$$\text{INR} = \frac{w^\dagger Rw}{w^\dagger w} \tag{2.8}$$

The adaptive array cancellation ratio, denoted C, is defined here as the ratio
of the summation of probe-received power before adaption to the summation
of probe-received power after adaption; that is,

$$C = \frac{p_q}{p_a} \tag{2.9}$$

Substituting (2.7) in (2.9) yields

$$C = \frac{w_q{}^\dagger Rw_q}{w_a{}^\dagger Rw_a} \tag{2.10}$$

Next, the covariance matrix defined by the elements in (2.2) or (2.3) is Hermitian (that is, $R = R^\dagger$), which, by the spectral theorem, can be decomposed in eigenspace [39] as

$$R = \sum_{k=1}^{M} \lambda_k e_k e_k^\dagger \tag{2.11}$$

where $\lambda_k, k = 1, 2, \cdots, M$ are the eigenvalues of R, and $e_k, k = 1, 2, \cdots, M$ are the associated eigenvectors of R. The interference covariance matrix eigenvalues $(\lambda_1, \lambda_2, \cdots, \lambda_M)$ are a convenient quantitative measure of the use of the adaptive array degrees of freedom [19-23]. A count of the number of covariance matrix eigenvalues above receiver noise (above unity based on (2.4)) represents the number of adaptive array degrees of freedom consumed. If there are M adaptive channels in the array, as long as $M - 1$ or fewer degrees of freedom are fully consumed, the array can form adaptive nulls. The amplitude spread between the largest and smallest eigenvalues is a quantitative measure of the dynamic range of the interference (hot spot) signals. The eigenvalues of the covariance matrix can be computed by solving $\det(R - \lambda I) = 0$, where *det* means determinant. The covariance matrix eigenvalues are proportional to power and in decibels can be computed as $\lambda_{k\mathrm{dB}} = 10 \log_{10} \lambda_k$.

An example of the use of the SMI algorithm is now considered. Consider Figure 2.2, which shows two coherent opposing plane waves (denoted by the electric fields $E_1(x)$ and $E_2(x)$) surrounding a target region with electric-field (E-field) probes located at opposing positions A and B, and a focal point at position F. In this example, it is desired to form electric-field nulls at positions A and B while maintaining the focused beam at position F. For this simple example, assume a lossless homogeneous medium of air; however, the

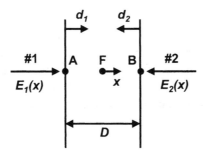

Figure 2.2 Diagram showing two coherent planes waves irradiating a target region of thickness D with a desired focus at position F.

analysis approach applies to an arbitrary medium. The covariance matrices associated with probe positions A and B are expressed in terms of the correlation of received probe voltages as

$$R_A = \begin{bmatrix} (V_{1A})(V_{1A})^* & (V_{1A})(V_{2A})^* \\ (V_{2A})(V_{1A})^* & (V_{2A})(V_{2A})^* \end{bmatrix} \tag{2.12}$$

$$R_B = \begin{bmatrix} (V_{1B})(V_{1B})^* & (V_{1B})(V_{2B})^* \\ (V_{2B})(V_{1B})^* & (V_{2B})(V_{2B})^* \end{bmatrix} \tag{2.13}$$

where V_{1A} and V_{2A} are the probe-received voltages at position A due to applicators 1 and 2, respectively. Similarly, V_{1B} and V_{2B} are the probe-received voltages at position B due to applicators 1 and 2, respectively. To compute the adaptive array weights, the inverse of the covariance matrix is needed. The inverse of a matrix, M, is given by

$$M^{-1} = \frac{\mathrm{adj}\,M}{\mathrm{det}\,M} \tag{2.14}$$

where *adj* means adjoint. Let M be expressed as

$$M = \begin{bmatrix} a & b \\ c & d \end{bmatrix} \tag{2.15}$$

The adjoint of matrix M is expressed as

$$\mathrm{adj}\,M = \begin{bmatrix} d & -b \\ -c & a \end{bmatrix} \tag{2.16}$$

and the determinant of matrix M is given by

$$\mathrm{det}\,M = ad - bc \tag{2.17}$$

Referring again to Figure 2.2, assume the case of plane-wave incidence from two coherent opposing applicators in a lossless medium, and assume that the thickness of the lossless medium is one-half wavelength, or $D = \lambda/2$. When a plane wave travels a distance of $\lambda/2$, the phase changes by $180°$. To compute the covariance matrix, assume that the coherent applicators radiate one at a time, and assume that the voltage received at the surface of the medium is proportional to the electric field E_o generated by the applicator on the corresponding side of the medium. Then the relative received voltages at E-field probes located at positions A and B are given by

$$V_{1A} = E_o \tag{2.18}$$

$$V_{1B} = -E_o \tag{2.19}$$

$$V_{2A} = -E_o \tag{2.20}$$

$$V_{2B} = E_o \tag{2.21}$$

where E_o is the incident electric field.

Substituting (2.18) and (2.20) in (2.12) yields

$$R_A = \begin{bmatrix} E_o^2 & -E_o^2 \\ -E_o^2 & E_o^2 \end{bmatrix} \tag{2.22}$$

and substituting (2.19) and (2.21) in (2.13) yields

$$R_B = \begin{bmatrix} E_o^2 & -E_o^2 \\ -E_o^2 & E_o^2 \end{bmatrix} \tag{2.23}$$

and so for this example it is observed that R_A is equal to R_B. From (2.4), the covariance matrix, R, is expressed as the summation of covariance matrices R_A and R_B as

$$R = R_A + R_B + I \tag{2.24}$$

Substituting (2.22) and (2.23) in (2.24) yields

$$R = \begin{bmatrix} 2E_o^2 + 1 & -2E_o^2 \\ -2E_o^2 & 2E_o^2 + 1 \end{bmatrix} \tag{2.25}$$

To compute the inverse of the covariance matrix, (2.14) is applied. It follows from (2.16) that the adjoint of R is given by

$$\text{adj}R = \begin{bmatrix} 2E_o^2 + 1 & 2E_o^2 \\ 2E_o^2 & 2E_o^2 + 1 \end{bmatrix} \tag{2.26}$$

and from (2.17) the determinant of R is given by

$$\det R = 4E_o^2 + 1 \tag{2.27}$$

from which the desired covariance matrix inverse follows.

In this simplified example, assume that the quiescent weight vector is given by

$$w_q = \begin{bmatrix} 1 \\ 1 \end{bmatrix} \tag{2.28}$$

that is, both applicators are radiating coherently and in phase and, hence, are centrally focused prior to adaptive nulling. However, it should be noted that in the two-channel case, the quiescent weight vector can always be chosen as

$(1, 0)^T$ to compute the adaptive array weights for any desired focal position or nulling position, and the same approach (one element is on and the rest are off) applies for more than two channels. The adaptive array weights can now be computed using (2.14), (2.26), (2.27), and (2.28), in (2.5) with the result

$$w_a = \begin{bmatrix} 1 \\ 1 \end{bmatrix} \qquad (2.29)$$

For this array geometry of opposing applicators, the centrally focused condition also corresponds to the weighting for adaptive nulling on the opposing surface sites. The first clinical application of adaptive phased array thermotherapy makes use of this inherent focusing and nulling characteristic of opposing microwave applicators for treating breast cancer as discussed in Chapters 8 and 9.

Now, continuing with this example, the received signal voltage vector at position A is denoted as V_A and from (2.18) and (2.20) is given by

$$V_A = \begin{bmatrix} V_{1A} \\ V_{2A} \end{bmatrix} = \begin{bmatrix} E_o \\ -E_o \end{bmatrix} \qquad (2.30)$$

The adaptive transmit phased array weighted received signal at position A can be expressed as

$$y_A = w_a^\dagger V_A \qquad (2.31)$$

Substituting (2.29) and (2.30) in (2.31), the reader can easily verify that the adaptive transmit phased array weighted received signal at position A, due to the opposing applicators, is

$$y_A = 0 \qquad (2.32)$$

Thus, a null is formed at position A when the coherent opposing applicators are driven in phase. By symmetry, a null is also formed at position B. At the same time, since the applicators are in phase and symmetrically located, there is a focused beam formed at the midpoint between the applicators. It is readily shown in the lossless free-space case that the normalized electric field between points A and B due to the two opposing plane waves can be expressed by superposition as

$$E(x) = E_1(x) + E_2(x) = E_o e^{-j\frac{2\pi}{\lambda}x} + E_o e^{j\frac{2\pi}{\lambda}x} = 2E_o \cos(\frac{2\pi}{\lambda}x) \quad (2.33)$$

and with $D = \lambda/2$ has a peak of $2E_o$ at $x = 0$ and a null at $x = \pm\lambda/4$. Power density is proportional to the electric-field magnitude squared (refer to Chapter 3, (3.61)). From (2.33), the focal power density due to the two coherent opposing plane waves is proportional to $4E_o^2$. For a single applicator, the power density at the focal position would be proportional to E_o^2. Thus, the focal power density of the coherent phased array is four times higher than for a single applicator. Note that the lossy case for two coherent opposing plane waves irradiating tissue will be addressed in Chapter 3.

In clinical practice, a single invasive microwave receiving probe can be positioned in tumor tissue to provide a focal feedback signal for adaptive focusing, as will be demonstrated in Chapter 9. In addition, E-field sensors can be positioned on the skin to null the surface fields. In the case of coherent opposing applicators, E-field sensors can be placed on opposing skin surfaces and the summation of probe-received power can be nullified to protect the skin during focused microwave thermotherapy treatment of cancer. The next section describes a gradient-search algorithm that can be used in clinical practice.

2.3 Gradient-Search Algorithm

Gradient-search algorithms are commonly used in adaptive array applications where the channel correlation is not readily calculated or measured. With a gradient search, only the output power of the receiver channels needs to be measured and is used as a feedback signal to the algorithm [24-38].

Under conditions where only the probe-received power is measured, it is appropriate to consider a gradient-search algorithm to minimize the E-field power at selected positions. The gradient search is used to control the transmit weights iteratively so that the RF signal received by the probe array is minimized. The transmit-array weights (gain and phase) are adaptively changed in small increments and the probe-array output power is monitored to determine the weight settings that reduce the output power most rapidly to a null. The mathematical formulation for the gradient search is developed in a straightforward manner [25, 37] and will now be described in the context of hyperthermia. Although the mathematical formulation is given as a minimization problem for adaptive nulling, the equations are readily converted to the maximization problem for adaptive focusing.

The summation of power received at the electric-field probes is denoted by p^{rec}. The adaptive array cancellation ratio, denoted C, is defined as before (in (2.9)) as the ratio of the summation of probe-received power before adaption p_q to the summation of probe-received power after adaption p_a; that is,

$$C = \frac{p_q}{p_a} \qquad (2.34)$$

Consider now J sets (or iterations) of N transmit weights that are applied to an adaptive hyperthermia phased array antenna. In terms of adaptive nulling, the optimum transmit-weight settings (from the collection of J sets of N transmit weights) occur when the total interference power received by the auxiliary probe array, denoted p^{rec}, is minimized. For notational convenience let a figure of merit F denote p^{rec} and employ a method of steepest-descent gradient search to find the optimum transmit weights to minimize F; that is,

$$F_{opt} = \min(F_j) \quad j = 1, 2, \cdots, J \tag{2.35}$$

Assume that there are N complex transmit weights in the hyperthermia phased array as suggested by the amplitude and phase scatter diagram depicted in Figure 2.3. The nth transmit weight in the jth configuration (or

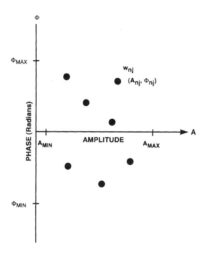

Figure 2.3 Scatter diagram in rectangular coordinates for the amplitude and phase of the transmit weights in a hyperthermia phased array antenna. The nth transmit weight in the jth configuration of transmit weights is denoted w_{nj}.

iteration) of transmit weights is denoted

$$w_{nj} = A_{nj}e^{j\Phi_{nj}} \tag{2.36}$$

where A_{nj} is the transmit-weight amplitude distributed over the range A_{min} to A_{max} and Φ_{nj} is the transmit-weight phase distributed over the range Φ_{min} to Φ_{max}. The goal is to find the values of amplitude and phase for each of the N transmit weights such that the figure of merit (p^{rec}) is minimized. When the figure of merit is minimized, adaptive radiation pattern nulls will be formed at the auxiliary sensor positions.

Assuming an initial setting of the N transmit weights, the weights are adjusted by dithering them until the optimum figure of merit is achieved. The goal is to find the collective search directions for the N transmit weights such that F decreases most rapidly; that is, the transmit weights are selected so that the directional derivative is minimized at (A_j, Φ_j), where A_j and Φ_j are the amplitude and phase column vectors, respectively.

The directional derivative of F_j is expressed in terms of the amplitude and phase changes of the transmit weights as

$$D(F_j) = \sum_{n=1}^{N} \left(\frac{\partial F_j}{\partial A_{nj}} r_{Anj} + \frac{\partial F_j}{\partial \Phi_{nj}} r_{\Phi nj} \right) \qquad (2.37)$$

where ∂ means partial derivative and $r_{Anj}, r_{\Phi nj}$ are the (A, Φ) directions for which F_j is decreasing most rapidly. The directions $r_{Anj}, r_{\Phi nj}$ are constrained by

$$\sum_{n=1}^{N} (r_{Anj}^2 + r_{\Phi nj}^2) = 1 \qquad (2.38)$$

The goal is to minimize $D(F_j)$ subject to the above constraint equation.

Using Lagrange multipliers, construct the Lagrangian function

$$L_j = \sum_{n=1}^{N} \left(\frac{\partial F_j}{\partial A_{nj}} r_{Anj} + \frac{\partial F_j}{\partial \Phi_{nj}} r_{\Phi nj} \right) + G[1 - \sum_{n=1}^{N} (r_{Anj}^2 + r_{\Phi nj}^2)] \qquad (2.39)$$

where G is a constant to be determined. The requirement that L_j be an extremum implies that the first derivatives of (2.39) be zero

$$\frac{\partial L_j}{\partial r_{Anj}} = \frac{\partial F_j}{\partial A_{nj}} - 2G r_{Anj} = 0, n = 1, 2, \cdots, N \qquad (2.40)$$

$$\frac{\partial L_j}{\partial r_{\Phi nj}} = \frac{\partial F_j}{\partial \Phi_{nj}} - 2G r_{\Phi nj} = 0, n = 1, 2, \cdots, N \qquad (2.41)$$

or solving (2.40) and (2.41)

$$r_{Anj} = \frac{1}{2G} \frac{\partial F_j}{\partial A_{nj}} \qquad (2.42)$$

$$r_{\Phi nj} = \frac{1}{2G} \frac{\partial F_j}{\partial \Phi_{nj}} \qquad (2.43)$$

Squaring (2.42) and (2.43) and invoking (2.38) yields

$$\sum_{n=1}^{N} (r_{Anj}^2 + r_{\Phi nj}^2) = 1 = \frac{1}{4G^2} \sum_{n=1}^{N} \left[\left(\frac{\partial F_j}{\partial A_{nj}} \right)^2 + \left(\frac{\partial F_j}{\partial \Phi_{nj}} \right)^2 \right] \qquad (2.44)$$

thus,

$$G = \pm \frac{1}{2} \sqrt{\sum_{n=1}^{N} [(\frac{\partial F_j}{\partial A_{nj}})^2 + (\frac{\partial F_j}{\partial \Phi_{nj}})^2]} \qquad (2.45)$$

Substituting (2.45) into (2.42) and (2.43) gives

$$r_{Anj} = -\frac{\frac{\partial F_j}{\partial A_{nj}}}{\sqrt{\sum_{n=1}^{N} [(\frac{\partial F_j}{\partial A_{nj}})^2 + (\frac{\partial F_j}{\partial \Phi_{nj}})^2]}} \qquad (2.46)$$

$$r_{\Phi nj} = -\frac{\frac{\partial F_j}{\partial \Phi_{nj}}}{\sqrt{\sum_{n=1}^{N} [(\frac{\partial F_j}{\partial A_{nj}})^2 + (\frac{\partial F_j}{\partial \Phi_{nj}})^2]}} \qquad (2.47)$$

In (2.46) and (2.47) the minus sign was chosen corresponding to the direction of maximum function decrease. (Note: by changing the minus sign to a plus sign in (2.46) and (2.47), the search directions then correspond to the direction of maximum function increase (method of steepest ascent; that is, the plus sign is used to maximize the power delivered to the focus or tumor site). The partial derivatives

$$\frac{\partial F_j}{\partial A_{nj}}, \text{ and } \frac{\partial F_j}{\partial \Phi_{nj}} \quad ; n = 1, 2, \cdots, N \qquad (2.48)$$

represent the gradient directions for maximum function decrease. An alternate figure of merit could be the ratio of the focal power to the nulling probe power that could be adaptively maximized.

In a hyperthermia treatment, because the figure of merit F is measured and cannot be expressed in analytical form, the partial derivatives are numerically evaluated using finite differences. Thus,

$$\frac{\partial F_j}{\partial A_{nj}} = \frac{\Delta F_{Anj}}{2\Delta A_{nj}} \qquad (2.49)$$

$$\frac{\partial F_j}{\partial \Phi_{nj}} = \frac{\Delta F_{\Phi nj}}{2\Delta \Phi_{nj}} \qquad (2.50)$$

where, as shown in Figure 2.4,

$$\Delta F_{Anj} = F_j(A_{nj} + \Delta A_{nj}; \Phi_{nj}) - F_j(A_{nj} - \Delta A_{nj}; \Phi_{nj}) \qquad (2.51)$$

$$\Delta F_{\Phi nj} = F_j(A_{nj}; \Phi_{nj} + \Delta \Phi_{nj}) - F_j(A_{nj}; \Phi_{nj} - \Delta \Phi_{nj}) \qquad (2.52)$$

and ΔA_{nj} and $\Delta \Phi_{nj}$ are the maximum step sizes. Assume for now that the increments ΔA_{nj} and $\Delta \Phi_{nj}$ depend on the iteration number j and transmit

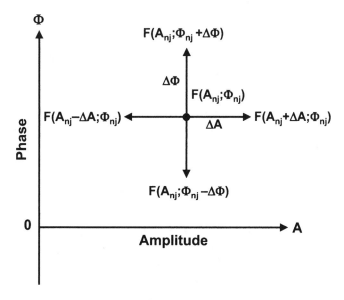

Figure 2.4 Figure of merit with transmit-weight dithering for optimum gradient-search directions.

element index n. Substituting (2.49), (2.50), (2.51), and (2.52) into (2.46) and (2.47) gives the desired result for the search directions

$$r_{Anj} = -\frac{\frac{\Delta F_{Anj}}{\Delta A_{nj}}}{\sqrt{\sum_{n=1}^{N}[(\frac{\Delta F_{Anj}}{\Delta A_{nj}})^2 + (\frac{\Delta F_{\Phi nj}}{\Delta \Phi_{nj}})^2]}} \quad (2.53)$$

$$r_{\Phi nj} = -\frac{\frac{\Delta F_{\Phi nj}}{\Delta \Phi_{nj}}}{\sqrt{\sum_{n=1}^{N}[(\frac{\Delta F_{Anj}}{\Delta A_{nj}})^2 + (\frac{\Delta F_{\Phi nj}}{\Delta \Phi_{nj}})^2]}} \quad (2.54)$$

The new amplitude and phase settings of the $(j+1)$th transmit-weight configuration are computed according to

$$A_{n,j+1} = A_{nj} + \Delta A_{nj} r_{Anj} \quad (2.55)$$

$$\Phi_{n,j+1} = \Phi_{nj} + \Delta \Phi_{nj} r_{\Phi nj} \quad (2.56)$$

It could be assumed (for convenience) that the step sizes are independent of both the iteration number and the adaptive channel number; that is,

$$\Delta A_{nj} = \Delta A \quad (2.57)$$

$$\Delta \Phi_{nj} = \Delta \Phi \qquad (2.58)$$

However, in some situations it may be desirable to change the step size at each iteration as considered by Farina and Flam [29], or as described below for fast acceleration and convergence of the adaptive phased array algorithm [8, 13, pp. 87-90].

The standard gradient search derived above can be slow in converging to the desired amplitude and phase settings to form a set of nulls or to focus the adaptive array. The reason that the standard gradient-search algorithm is slow is due to the number of measurements required at each iteration. At each iteration, four measurements are made as a result of the phase and amplitude dithering, and these measurements are used to determine the search direction. An improvement in convergence can be obtained by utilizing subiterations that make use of the current search directions to increment the amplitude and phase weights. In this way, the weights are updated in an accelerated fashion while the null power or focal power is monitored as described below.

To speed the convergence of the gradient search, the following modifications of (2.55) and (2.56) are used. The fast-acceleration amplitude and phase settings of the current jth transmit-weight configuration are computed by introducing subiterations denoted by the index k, where $k = 1, 2, \cdots, K$ such that

$$A_{n,j,k} = A_{nj} + \Delta A_{nj} r_{Anj} 2^{k-1} \qquad (2.59)$$

$$\Phi_{n,j,k} = \Phi_{nj} + \Delta \Phi_{nj} r_{\Phi nj} 2^{k-1} \qquad (2.60)$$

In other words, at each iteration j, the algorithm starts a subiteration k that changes the amplitude and phase increments in increasing powers of 2. Note that other values besides 2 can be used, including smaller values and larger values. When the subiteration is started, $k = 1$ and the adaptive array weights $A_{n,j,1}$ and $\Phi_{n,j,1}$ are calculated and can be set via digital to analog converters in the hardware. The electric-field probe powers $p_{j,k,i}^{rec}$, $i = 1, 2, \cdots, N_{aux}$ at iteration j and subiteration k are measured and stored in the computer. In the case of adaptive nulling, the algorithm can be made to stop when either the individual electric-field probe-received powers reach the desired relative null-strength values or when the summation of the electric-field probe powers reach the desired relative null-strength value. For adaptive focusing, the algorithm can stop when no further focusing (no further increase in focal power) is achieved. During the next subiteration, $k = 2$, the adaptive array weights $A_{n,j,2}$ and $\Phi_{n,j,2}$ are computed according to (2.59) and (2.60). These new weights are set by the hardware and the probe-received powers $p_{j,k,i}^{rec}$, $i = 1, 2, \cdots, N_{aux}$ at iteration j and subiteration $k = 2$ are measured and stored in the computer.

Let it be assumed here that in the case of adaptive nulling, the desired figure of merit is equal to the summation of received-probe powers that is to be minimized. For adaptive nulling, if

$$\sum_{i=1}^{N_{aux}} p_{j,k=2,i}^{rec} < \sum_{i=1}^{N_{aux}} p_{j,k=1,i}^{rec} \qquad (2.61)$$

then the summation of received-probe power has decreased and the subiterations continue by incrementing k to 3 and proceeding in the same manner. That is, compute and set $A_{n,j,3}$ and $\Phi_{n,j,3}$, measure the received-probe powers, and compare the magnitude of $\sum_{i=1}^{N_{aux}} p_{j,k=3,i}^{rec}$ with $\sum_{i=1}^{N_{aux}} p_{j,k=2,i}^{rec}$ as in the previous subiteration. However, for adaptive nulling if

$$\sum_{i=1}^{N_{aux}} p_{j,k=2,i}^{rec} > \sum_{i=1}^{N_{aux}} p_{j,k=1,i}^{rec} \qquad (2.62)$$

then the summation of received-probe power has increased and the subiterations stop and the next iteration for j continues. The step sizes ΔA_{nj} and $\Delta \Phi_{nj}$ can be constant at each subiteration or they can be adjusted as desired.

Figure 2.5 shows a block diagram of an adaptive phased array hyperthermia system controlled by the fast-acceleration gradient-search algorithm. Each of the transmit phased array antenna elements induces a voltage across the terminals of the ith receive field probe antenna. For any given configuration of the transmit weights, each weight is dithered by a small amount in amplitude and phase, and the received powers at the E-field probes are stored in the computer for calculation of the figure of merit, gradient-search directions, and updated transmit-weight configuration. One transmit weight is dithered with the remaining transmit weights in their state prior to dithering. The figure of merit at the jth iteration and kth subiteration, denoted $F_{j,k}$, in the adaptive hyperthermia system is the power received by the auxiliary probe array, as indicated in the block diagram. Search directions for the adaptive transmit weights are based on minimizing the auxiliary probe array received power and are computed based on (2.53) and (2.54). Transmit weights for the next configuration $(j + 1)$ are computed from (2.59) and (2.60). The adaptive weight vector w_a is achieved when the figure of merit $F_{j,k}$ has converged.

2.4 SUMMARY

An adaptive receive or transmit phased array system relies on feedback signals as well as a computerized algorithm that can determine the desired amplitude

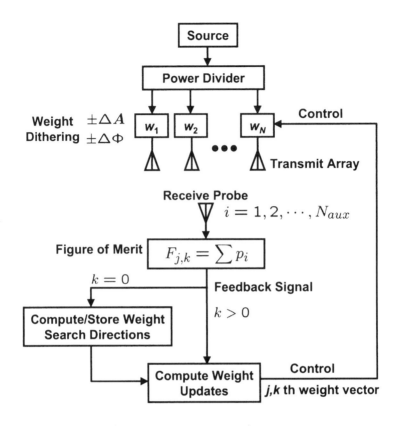

Figure 2.5 Fast-acceleration gradient-search algorithm block diagram for an adaptive phased array hyperthermia system.

and phase settings for each antenna element of the array, such that a desired radiation pattern is achieved. For an adaptive transmit phased array system for cancer treatment, the feedback signals are provided by probe antenna sensors on or in the target body. Two example adaptive array algorithms that apply for radar, communications, and hyperthermia treatment of cancer applications are the sample matrix inversion algorithm (SMI) and the gradient-search (GS) algorithm. In this chapter, the sample matrix inversion and gradient-search algorithms were described in the context of an adaptive transmitting phased array antenna such as applied to treating cancer. The next chapter provides a detailed background for electromagnetic field theory for tissue heating.

2.5 PROBLEM SET

2.1 For the quiescent weight vector given by (2.28) and the covariance matrix given by (2.25), verify that the adaptive weight vector is given by (2.29).

2.2 Assume a quiescent weight vector given by $w_q = (1,0)^T$ and the covariance matrix given by (2.25) with $E_o^2 = 100$. Show that the adaptive weight vector using (2.5) is $w_a \approx (1,1)^T$. Note that the value of $E_o^2 = 100$ is equivalent to a 20-dB signal-to-noise ratio.

2.3 Assume a quiescent weight vector is given by $w_q = (1, w_2)^T$, where w_2 is an arbitrary complex weight constrained by $|w_2| \leq 1$. Show that for the case of $E_o^2 >> 1$ (large signal-to-noise ratio) that the adaptive weight vector computed using the covariance matrix given by (2.25) is $w_a = (1,1)^T$.

2.4 Compute the covariance matrix eigenvalues in decibels from (2.25) using $E_o = 10$. Hint: Solve $\det(R - \lambda I) = 0$ for λ_1 and λ_2 in decibels relative to receiver noise. Ans. $\lambda_1 = 26.0$ dB, $\lambda_2 = 0.0$ dB. For this two-element adaptive array, there is only one large eigenvalue relative to receiver noise so one degree of freedom (out of two total available) is consumed. Thus, since not all of the degrees of freedom are consumed the adaptive array can form nulls at the desired nulling positions.

2.5 Verify the result given by (2.32); that is, the adaptive received signal amplitude on the opposing surface (denoted as point A) is zero. Repeat for surface position B.

References

[1] Fenn, A.J., "Adaptive Nulling Hyperthermia Array," US Patent No. 5,251,645, October 12, 1993.

[2] Fenn, A.J., "Adaptive Focusing and Nulling Hyperthermia Annular and Monopole Phased Array Applicators," US Patent No. 5,441,532. August 15, 1995.

[3] Fenn, A.J., and G.A. King, "Experimental Investigation of an Adaptive Feedback Algorithm for Hot Spot Reduction in Radio-Frequency Phased-Array Hyperthermia," *IEEE Trans Biomed Eng.*, Vol. 43, No. 3, 1994, pp. 273-280.

[4] Fenn, A.J., and G.A. King, "Adaptive Radio Frequency Hyperthermia Phased Array System for Improved Cancer Therapy: Phantom Target Measurements," *Int J Hyperthermia*, Vol. 10, No. 2, 1994, pp. 189-208.

[5] Fenn, A.J., "Minimally Invasive Monopole Phased Array Hyperthermia Applicators and Method for Treating Breast Carcinomas," US Patent No. 5,540,737, July 30, 1996.

[6] Sathiaseelan, V., A.J. Fenn, and A. Taflove, "Recent Advances in External Electromagnetic Hyperthermia," In: Chapter 10 of *Advances in Radiation Treatment*, Mittal, B.B., J.A. Purdy, and K.K. Ang, (eds.), Boston, Massachusetts: Kluwer Academic Publishers, 1998, pp. 213-245.

[7]   Fenn, A.J., V. Sathiaseelan, G.A. King, and P.R. Stauffer, "Improved Localization of Energy Deposition in Adaptive Phased Array Hyperthermia Treatment of Cancer," *J Oncol Management*, Vol. 7, No. 2, 1998, pp. 22-29.

[8]   Fenn, A.J., "Thermodynamic Adaptive Phased Array System for Activating Thermosensitive Liposomes in Targeted Drug Delivery," US Patent No. 5,810,888, September 22, 1998.

[9]   Fenn, A.J., G.L. Wolf, and R.M. Fogle, "An Adaptive Phased Array for Targeted Heating of Deep Tumors in Intact Breast: Animal Study Results," *Int J Hyperthermia*, Vol. 15, No. 1, 1999, pp. 45-61.

[10] Gavrilov, L.R., J.W. Hand, J.W. Hopewell, and A.J. Fenn, "Pre-clinical Evaluation of a Two-Channel Microwave Hyperthermia System with Adaptive Phase Control in a Large Animal," *Int J Hyperthermia*, Vol. 15, No. 6, 1999, pp. 495-507.

[11] Gardner, R.A., H.I. Vargas, J.B. Block, C.L. Vogel, A.J. Fenn, G.V. Kuehl, and M. Doval, "Focused Microwave Phased Array Thermotherapy for Primary Breast Cancer," *Ann Surg Oncol*, Vol. 9, No. 4, 2002, pp. 326-332.

[12] Vargas, H.I., W.C. Dooley, R.A. Gardner, K.D. Gonzalez, S.H. Heywang-Kobrunner, and A.J. Fenn, "Focused Microwave Phased Array Thermotherapy for Ablation of Early-Stage Breast Cancer: Results of Thermal Dose Escalation," *Ann Surg Oncol*, Vol. 11, No. 2, 2004, pp. 139-146.

[13] Fenn, A.J., *Breast Cancer Treatment by Focused Microwave Thermotherapy*, Sudbury, MA: Jones and Bàrtlett, 2007.

[14] Vargas, H.I., W.C. Dooley, A.J. Fenn, M.B. Tomaselli, and J.K. Harness, "Study of Preoperative Focused Microwave Phased Array Thermotherapy in Combination with Neoadjuvant Anthracycline-Based Chemotherapy for Large Breast Carcinomas," *Cancer Therapy*, Vol. 5, 2007, pp. 401-408, published online (www.cancer-therapy.org), November 25, 2007.

[15] Brennan, L.E., and I.S. Reed, "Theory of Adaptive Radar," *IEEE Trans. Aerospace and Electronic Systems,* Vol. 9, No. 2, 1973, pp. 237-252.

[16] Reed, I.S., J.D. Mallet, and L.E. Brennan, "Rapid Convergence Rate in Adaptive Arrays," *IEEE Trans. Aerospace and Electronic Systems*, Vol. AES-10, No. 6, Nov. 1974, pp. 853-863.

[17] Boroson, D.M., "Sample Size Considerations for Adaptive Arrays," *IEEE Trans. Aeorospace and Electronic Systems*, Vol. AES-16, No. 4, July 1980, pp. 446-451.

[18] Johnson, J.R., A.J. Fenn, H.M. Aumann, and F.G. Willwerth, "An Experimental Adaptive Nulling Receiver Utilizing the Sample Matrix Inversion Algorithm with Channel Equalization," *IEEE Trans. Microwave Theory Techniques*, Vol. 39, No. 5, 1991, pp. 798-808.

[19] Howells, P.W., "Explorations in Fixed and Adaptive Resolution at GE and SURC," *IEEE Trans. Antennas Propagat.*, Vol. 24, No. 5, 1976, pp. 575-584.

[20] Applebaum, S.P, "Adaptive Arrays," *IEEE Trans. Antennas Propagat.*, Vol. 24, No. 5, 1976, pp. 585-598.

[21] Gabriel, W.F., "Adaptive Arrays - An Introduction," *Proc. IEEE,* Vol. 64, 1976, pp. 239-271.

[22] Mayhan, J.T., "Some Techniques for Evaluating the Bandwidth Characteristics of Adaptive Nulling Systems," *IEEE Trans. Antennas Propagat.,* Vol. 27, No. 3, 1979, pp. 363-373.

[23] Mayhan, J.T., A.J. Simmons, and W.C. Cummings, "Wide-Band Adaptive Antenna Nulling Using Tapped Delay Lines," *IEEE Trans. Antennas Propagat.,* Vol. 29, No. 6, 1981, pp. 923-936.

[24] Monzingo, R.A., and T. W. Miller, *Introduction to Adaptive Arrays*, New York: Wiley. 1980.

[25] Fenn, A.J., "Maximizing Jammer Effectiveness for Evaluating the Performance of Adaptive Nulling Array Antennas," *IEEE Trans. Antennas Propagat.*, Vol. 33, No. 10, 1985, pp. 1131-1142.

[26] Chan, V.W.S., "A Fast Algorithm for Spatial Interference Rejection," *Int Conf on Communications*, Vol. 1, June 1980, pp. 59.4.1-59.4.7.

[27] Cantoni, A. "Application of Orthogonal Perturbation Sequences to Adaptive Beamforming," *IEEE Trans. Antennas Propagat.*, Vol. 28, No. 2, 1980, pp. 191-202.

[28] Farden, D.C., and R.M. Davis, "Orthogonal Weight Perturbation Algorithms in Partially Adaptive Arrays," *IEEE Trans. Antennas Propagat.* Vol. AP-33, No. 1, 1985, pp. 56-63.

[29] Farina, D.J., and R.P. Flam, "A Self-Normalizing Gradient-Search Adaptive Array Algorithm," *IEEE Trans Aerospace and Electronic Systems,* Vol. 27, No. 6, 1991, pp. 901-905.

[30] Godara, L.C., "Performance Analysis of Structured Gradient Algorithm," *IEEE Trans Antennas and Propagation,* Vol. 38, No. 7, 1990, pp. 1078-1083.

[31] Compton, Jr., R.T., *Adaptive Antennas, Concepts and Performance*, Upper Saddle River, NJ: Prentice-Hall, 1988.

[32] Chandran, S., (ed.), *Adaptive Antenna Arrays: Trends and Applications*, Berlin: Springer-Verlag, 2004.

[33] Nitzberg, R., *Adaptive Signal Processing for Radar*, Norwood, MA: Artech House, 1992.

[34] Farina, A., *Antenna-Based Signal Processing Techniques for Radar Systems,* Norwood, MA: Artech House, 1992.

[35] Hudson, J.E., *Adaptive Array Principles*, New York: Peter Peregrinus LTD, 1981.

[36] Fenn, A.J., *Adaptive Antennas and Phased Arrays for Radar and Communications*, Norwood, MA: Artech House, 2008.

[37] Zahradnik, R.L., *Theory and Techniques of Optimization for Practicing Engineers,* New York: Barnes and Noble, New York, 1971, pp. 118-124.

[38] Hasdorff, L., *Gradient Optimization and Nonlinear Control,* New York: John Wiley & Sons, 1976.

[39] Strang, G., *Linear Algebra and Its Applications,* New York: Academic, 1976, pp. 211-227.

Free Subscription
Artech Direct email newsletter

New Title News • Special Offers • Author Insights

☐ Yes! Please enter my free subscription to *Artech Direct* and keep me up-to-date with emailed news of product and service information from Artech House/Horizon House Publishers.

email address: _____

You may also make my email address available to selected industry organizations and companies. ☐ Yes ☐ No

Please indicate your areas of interest

☐ Telecommunications/Wireless/Networking

☐ Software Engineering/Computer Security

☐ Microwave

☐ Radar/Remote Sensing/Electronic Defense

☐ Signal Processing

☐ Sensors/MEMS/Nanotechnology

☐ Antennas & Propagation

☐ Engineering Management

☐ Biomedical Engineering

Mailing address:

Name: _____

Company: _____

Address: _____

_____

Fax or mail this card to the Artech House office nearest you. Please see other side.

 **ARTECH HOUSE** BOSTON | LONDON

To see the full line of Artech House books and software, visit our online bookstore at:

www.artechhouse.com

Special Offers • Sample Chapters • Secure Ordering

To receive information on new and forthcoming titles from Artech House, please fill out the other side of this card and mail or fax it to one of the locations below:

For Europe, Asia, Middle East, Africa:

Artech House
46 Gillingham Street
London, SW1V 1AH U.K.
+44 (0)20 7596 8750
FAX: +44 (0)20 7630 0166
artech-uk@artechhouse.com

All other regions:

Artech House
685 Canton Street
Norwood, MA 02062 U.S.A.
1-781-769-9750
1-800-225-9977 (continental U.S. only)
FAX: 1-781-769-6334
artech@artechhouse.com

ARTECH HOUSE BOSTON | LONDON

3

Electromagnetic Field Theory for Tissue Heating

3.1 INTRODUCTION

Radiofrequency or microwave electromagnetic fields can induce a temperature increase in tissue provided that the power density of the electromagnetic field is sufficiently strong [1, pp. 518-520], [2, pp. 19-22], [3, p. 217]. Due to water and ion content, human tissues are fundamentally lossy in terms of electromagnetic wave propagation, attenuation, and absorption [4, pp. 167-223]. The absorbed electromagnetic energy gives rise to an elevated temperature in the tissue.

For cancer treatment, ideally the electromagnetic energy should be deposited in tumor tissue with minimal energy deposited in tissue that is free of cancer cells. The water content of human tissues varies considerably as shown in Table 3.1 for data presented by Woodard and White [5]. For example, adipose tissue (often referred to in the literature as fatty tissue) has a water content in the range of about 10% to 30%, which is considered low. Normal breast tissue has a water content in the range of about 30% to 70%, which can be considered low to moderate. Connective tissue has a moderate water content on the order of 60%. Muscle tissue has a high water content in the range of about 75% to 80%. The water content of the prostate gland is about 83%, which is high. Cancerous tissue tends to have high water content as shown for the tumors listed at the bottom of Table 3.1. Breast carcinoma has a water content in the range of 66% to 79%, liver cancer 82%, and skin cancer 82%. Body tissues are composed of various amounts of water, lipid (long-chain fatty acids), protein, carbohydrates, and minerals [5]. An atom or a group of atoms having a net positive or negative charge is known as an ion.

Table 3.1

Approximate Water Content of Human Tissues as a Percentage of Mass

Tissue	Water Content, Percent by Mass
Adipose	10% to 30%
Bladder	70% to 79%
Blood plasma	92%
Blood whole	79%
Bone (cortical)	12%
Brain (grey matter)	83%
Brain (white matter)	69%
Breast	30% to 73%
Connective tissue	60%
Esophagus	70% to 79%
Heart	71% to 81%
Intestine (small)	81%
Kidney	72% to 81%
Liver	73% to 76%
Lung	80%
Lymph	92%
Muscle	70% to 79%
Ovary	83%
Pancreas	73%
Prostate	83%
Skin	59% to 72%
Tumor (Breast cancer)	66% to 79%
Tumor (Liver cancer)	82%
Tumor (Skin cancer)	82%

Source: normal tissues, Woodard and White [5].
Source: malignant breast, Campbell and Land [7].
Source: malignant liver and skin, Foster and Schepps [8].

Cations are positively charged ions and anions have a negative charge. Some examples of ions in tissues include calcium (Ca^{2+}), chloride (Cl^-), oxide (O^{-2}), phosphide (P^{-3}), potassium (K^+), sodium (Na^+), and sulfide (S^{-2}). Table 3.2 lists the elemental composition of human tissues. It is observed that normal breast tissue has a low ion content compared to muscle tissue for the ions O^{-2}, P^{-3}, S^{-2}, and K^+.

In radiofrequency and microwave hyperthermia, the electric-field component is a polarized vector quantity that applies a force to dipole molecules such as water and to charges or ions present in the tissue as described by von Hippel [6, p. 7]. When the electric field is time-varying in tissue, the applied force affects the water molecules and ions and they rotate, move, and collide

Table 3.2

Elemental Composition (by Percent Mass) for Body Tissues

Tissue	Elemental Composition (by Percent Mass)									
	H	C	N	O	Na	P	S	Cl	K	Ca
Adipose	11.4	59.8	0.7	27.8	0.1		0.1	0.1		
Bladder (empty)	10.5	9.6	2.6	76.1	0.2	0.2	0.2	0.3	0.3	
Blood (plasma)	10.8	4.1	1.1	83.2	0.3		0.1	0.4		
Blood (whole)	10.2	11.0	3.3	74.5	0.1	0.1	0.2	0.3	0.2	
Bone (cortical)	3.4	15.5	4.2	43.5	0.1	10.3	0.3			22.5
Brain (grey)	10.7	9.5	1.8	76.7	0.2	0.3	0.2	0.3	0.3	
Brain (white)	10.6	19.4	2.5	66.1	0.2	0.4	0.2	0.3	0.3	
Breast	10.6	33.2	3.0	52.7	0.1	0.1	0.2	0.1		
Connective	9.4	20.7	6.2	62.2	0.6		0.6	0.3		
Heart	10.3	12.1	3.2	73.4	0.1	0.1	0.2	0.3	0.2	
Kidney	10.3	13.2	3.0	72.4	0.2	0.2	0.2	0.2	0.2	0.1
Liver	10.2	13.9	3.0	71.6	0.2	0.3	0.3	0.2	0.3	
Lung	10.3	10.5	3.1	74.9	0.2	0.2	0.3	0.3	0.2	
Muscle	10.2	14.3	3.4	71.0	0.1	0.2	0.3	0.1	0.4	
Ovary	10.5	9.3	2.4	76.8	0.2	0.2	0.2	0.2	0.2	
Pancreas	10.6	16.9	2.2	69.4	0.2	0.2	0.1	0.2	0.2	
Prostate	10.5	8.9	2.5	77.4	0.2	0.1	0.2		0.2	
Skin	10.0	20.4	4.2	64.5	0.2	0.1	0.2	0.3	0.1	
Protein	6.6	53.4	17.0	22.0			1.0			
Water	11.2			88.8						

Source: Woodard and White [5].

with each other. The net molecular and ionic collisions produce heat energy, which in turn elevates the tissue temperature.

As will be discussed in Section 3.2, both the amplitude and phase of a propagating electromagnetic wave are affected by the dielectric properties of the various tissues and boundaries between tissues [1-4, 6-18]. Water and ion content of the tissue affect the dielectric properties. The dielectric properties of water and saline are well characterized [19, 20]. For hyperthermia treatment of cancer, the electromagnetic field propagation in tissue can be described in some cases as a spherical wave, or a collection of spherical waves. The electromagnetic wave can be generated by a single antenna or multiple antennas such as in a phased array or an adaptive phased array [21] [22, pp. 53-65]. For the case of an array applicator, each of the propagating waves will be partially reflected and partially transmitted at each type of tissue encountered by the propagating wave(s). Figure 3.1 shows an example case where an electromagnetic wave is incident on a tissue boundary that separates medium 1 from medium 2. The general electrical parameters that characterize

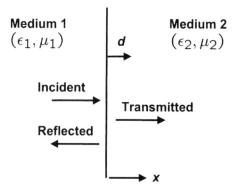

Figure 3.1 General case of an electromagnetic wave incident on a tissue interface. A portion of the incident wave from medium 1 described by dielectric parameters ϵ_1 and μ_1 will be reflected, and a portion will be transmitted into medium 2 described by dielectric parameters ϵ_2 and μ_2.

medium 1 and medium 2 are the permittivity ϵ and permeability μ. In the case where medium 2 is composed of electromagnetic lossy tissue, the wave that is transmitted will be attenuated versus depth d in the tissue.

This chapter is organized as follows. The next section describes the theory for an electromagnetic wave propagating in a conducting medium corresponding to lossy tissue. A brief discussion of the radiofrequency and microwave properties of tissues then follows. The subsequent sections then describe the mathematical formulation for electromagnetic analysis in homogeneous tissue by the method of moments and the finite-difference time-domain method for analysis of heterogeneous tissues.

3.2 WAVE PROPAGATION IN CONDUCTING MEDIA

3.2.1 MAXWELL'S ELECTROMAGNETIC FIELD EQUATIONS

To gain insight into the effect of lossy tissue on the propagation of an electromagnetic wave, it is useful to review certain fundamental equations that govern the field characteristics. The lossy (conducting) medium considered here is characterized by a complex permittivity, denoted ϵ_c and a real permeability μ.

As described in many textbooks on electromagnetic theory, electromagnetic fields can be analyzed by Maxwell's equations [23-39]. An overview of the development of Maxwell's equations is given in [24, pp. 45-55]. In time-dependent differential form, in a conducting medium Maxwell's equations are given by

$$\nabla \times \boldsymbol{H} = \boldsymbol{J} + \epsilon' \frac{\partial \boldsymbol{E}}{\partial t} \tag{3.1}$$

$$\nabla \times \boldsymbol{E} = -\mu \frac{\partial \boldsymbol{H}}{\partial t} \tag{3.2}$$

$$\nabla \cdot \boldsymbol{D} = \rho \tag{3.3}$$

$$\nabla \cdot \boldsymbol{B} = 0 \tag{3.4}$$

where $\boldsymbol{E}$ is the vector electric field with units of volts/meter, $\boldsymbol{H}$ is the vector magnetic field with units of amperes/meter, $\boldsymbol{J}$ is the conduction current density with units of amperes/meters^2, ϵ' is the real part of the complex permittivity of the medium, ρ is the volume electric charge density with units of coulomb/meter^3, $\boldsymbol{D} = \epsilon' \boldsymbol{E}$ is the electric flux density with units of coulombs/meter^2, $\boldsymbol{B} = \mu \boldsymbol{H}$ is the magnetic flux density with units of webers/meter^2, $\nabla \times$ is the curl operator, and $\nabla \cdot$ is the divergence operator. Bold lettering or bold symbols are used here to represent vectors or vector operations. An overview for computational methods for solving Maxwell's equations for the electromagnetic fields used in hyperthermia is given in [1, pp. 305-343] and [2, pp. 13-16].

The instantaneous electric and magnetic fields, as a function of spatial position (x, y, z) and time t, are real functions, and for the time-harmonic case, they are sinusoidal (or cosinusoidal) and can be expressed as

$$\boldsymbol{E}(x, y, z, t) = |\boldsymbol{E}(x, y, z)| \cos(\omega t + \psi_e(x, y, z)) \tag{3.5}$$

$$\boldsymbol{H}(x, y, z, t) = |\boldsymbol{H}(x, y, z)| \cos(\omega t + \psi_h(x, y, z)) \tag{3.6}$$

where $\psi_e(x, y, z)$ and $\psi_h(x, y, z)$ are the spatial phase variations of the electric and magnetic fields, respectively, and $\omega = 2\pi f$ is the radian frequency, where f is the frequency. Equivalently, the instantaneous electric and magnetic fields can be written in a more convenient exponential form in terms of the nontime-varying complex electric and magnetic fields $\boldsymbol{E}(x, y, z)$ and $\boldsymbol{H}(x, y, z)$ as

$$\boldsymbol{E}(x, y, z, t) = \text{Re}[\boldsymbol{E}(x, y, z)e^{j\omega t}] \tag{3.7}$$

$$\boldsymbol{H}(x, y, z, t) = \text{Re}[\boldsymbol{H}(x, y, z)e^{j\omega t}] \tag{3.8}$$

and similarly, the instantaneous conduction current density $\boldsymbol{J}$ can be written as

$$\boldsymbol{J}(x, y, z, t) = \text{Re}[\boldsymbol{J}(x, y, z)e^{j\omega t}] \tag{3.9}$$

where $\text{Re}[\cdot]$ indicates the real part.

The time-dependent fields can also be expressed in terms of the magnitude and phase of the fields as

$$\boldsymbol{E}(x, y, z, t) = \text{Re}[|\boldsymbol{E}(x, y, z)|e^{j(\omega t + \psi_e(x,y,z))}] \tag{3.10}$$

$$\boldsymbol{H}(x, y, z, t) = \text{Re}[|\boldsymbol{H}(x, y, z)|e^{j(\omega t + \psi_h(x,y,z))}] \tag{3.11}$$

If the $\text{Re}[\cdot]$ operation in (3.10) and (3.11) is dropped and the $e^{j\omega t}$ factor is suppressed, then the following phasor designations for the electric and magnetic fields can also be used

$$\boldsymbol{E}(x, y, z) = |\boldsymbol{E}(x, y, z)|e^{j\psi_e(x,y,z)} \tag{3.12}$$

$$\boldsymbol{H}(x, y, z) = |\boldsymbol{H}(x, y, z)|e^{j\psi_h(x,y,z)} \tag{3.13}$$

With the above relations, it is now possible to convert Maxwell's equations to a time-harmonic form as follows. Substituting (3.7), (3.8), and (3.9) into (3.1) and (3.2) yields

$$\boldsymbol{\nabla} \times \text{Re}[\boldsymbol{H}(x, y, z)e^{j\omega t}] = \text{Re}[\boldsymbol{J}(x, y, z)e^{j\omega t}] + \epsilon' \frac{\partial}{\partial t}\text{Re}[\boldsymbol{E}(x, y, z)e^{j\omega t}] \tag{3.14}$$

$$\boldsymbol{\nabla} \times \text{Re}[\boldsymbol{E}(x, y, z)e^{j\omega t}] = -\mu \frac{\partial}{\partial t}\text{Re}[\boldsymbol{H}(x, y, z)e^{j\omega t}] \tag{3.15}$$

and computing the partial derivatives yields

$$\boldsymbol{\nabla} \times \text{Re}[\boldsymbol{H}(x, y, z)e^{j\omega t}] = \text{Re}[\boldsymbol{J}(x, y, z)e^{j\omega t}] + j\omega\epsilon'\text{Re}[\boldsymbol{E}(x, y, z)e^{j\omega t}] \tag{3.16}$$

$$\boldsymbol{\nabla} \times \text{Re}[\boldsymbol{E}(x, y, z)e^{j\omega t}] = -j\omega\mu\text{Re}[\boldsymbol{H}(x, y, z)e^{j\omega t}] \tag{3.17}$$

The above two equations can be simplified by dropping both the real (Re) operation and suppressing the exponential function $e^{j\omega t}$ that are common to all terms, and further dropping the position coordinates (x, y, z) with the result for Maxwell's equations in time-harmonic (frequency domain) form

$$\boldsymbol{\nabla} \times \boldsymbol{H} = \boldsymbol{J} + j\omega\epsilon'\boldsymbol{E} \tag{3.18}$$

$$\boldsymbol{\nabla} \times \boldsymbol{E} = -j\omega\mu\boldsymbol{H} \tag{3.19}$$

$$\boldsymbol{\nabla} \cdot \boldsymbol{D} = \rho \tag{3.20}$$

$$\boldsymbol{\nabla} \cdot \boldsymbol{B} = 0 \tag{3.21}$$

Since tissues are principally nonmagnetic (blood, however, does have small amounts of iron [5]), in Maxwell's equations, the permeability μ can be expressed as $\mu = \mu_r\mu_o$, where μ_r is the relative permeability and μ_o is the permeability of free space. The real part of the complex permittivity is expressed as

$$\epsilon' = \epsilon'_r \epsilon_o \tag{3.22}$$

where ϵ'_r is the dielectric constant (relative permittivity) and ϵ_o is the permittivity of free space. In free space, the speed of light c (also the speed of electromagnetic wave propagation) is related to ϵ_o and μ_o as

$$c = \frac{1}{\sqrt{\epsilon_o \mu_o}} \tag{3.23}$$

and has a measured value of approximately $2.9979 \times 10^8 \approx 3.0 \times 10^8$ m/s. The value of ϵ_o can be measured by an experiment involving Coulomb's force law for two charges separated by a distance [24, p. 61]. The established value for the permittivity of free space is $\epsilon_o = \frac{1}{36\pi} \times 10^{-9} = 8.8542 \times 10^{-12}$ farads/meter (farads have units of amperes-seconds/volts). The established value for the permeability of free space is $\mu_o = 4\pi \times 10^{-7} = 1.257 \times 10^{-6}$ henries/meter (henries have units of volts-seconds/ampere).

For a conducting medium with electrical conductivity σ having units of siemens/meter (amperes per volt-meter), J and E are related as

$$J = \sigma E \tag{3.24}$$

which is Ohm's law for conducting current density. Note that electrical conductivity is equal to the reciprocal of electrical resistivity. Substituting (3.24) into (3.18) yields

$$\nabla \times H = (\sigma + j\omega\epsilon')E \tag{3.25}$$

The electrical conductivity σ of the dielectric medium can be expressed in terms of the imaginary component (denoted ϵ'') of the complex permittivity as

$$\sigma = \omega\epsilon'' = \omega\epsilon_o\epsilon''_r \tag{3.26}$$

where ϵ''_r is the relative imaginary component of the compex permittivity. Thus, using (3.22) and (3.26) it follows that (3.25) can be expressed as

$$\nabla \times H = (j\omega\epsilon_o\epsilon'_r + \omega\epsilon_o\epsilon''_r)E \tag{3.27}$$

which can be written in an alternate form (by factoring the quantity $j\omega\epsilon_o$) as,

$$\nabla \times H = j\omega\epsilon_o(\epsilon'_r - j\epsilon''_r)E \tag{3.28}$$

In (3.28), the complex permittivity of the conducting medium can be expressed as

$$\epsilon_c = \epsilon_o(\epsilon'_r - j\epsilon''_r) \tag{3.29}$$

Thus, (3.28) can be written in a simplified form in terms of the complex permittivity given by (3.29) as,

$$\nabla \times H = j\omega\epsilon_c E \tag{3.30}$$

3.2.2 VECTOR WAVE EQUATION

From (3.19) and (3.25), the vector wave equation in terms of E is derived as (see Problem 3.7)

$$\nabla^2 E - \gamma^2 E = 0 \tag{3.31}$$

where γ is the complex propagation constant. A similar vector wave equation can be derived for the magnetic field component (Problem 3.8). The vector wave equation allows for spherical, cylindrical, and plane wave solutions. It is readily shown that the propagation constant in (3.31) can be expressed as

$$\gamma = \pm\sqrt{j\omega\mu(\sigma + j\omega\epsilon')} = \pm j\omega\sqrt{\mu\epsilon'}\sqrt{1 - j\frac{\sigma}{\omega\epsilon'}} \tag{3.32}$$

The quantity $\sigma/\omega\epsilon' = \epsilon''/\epsilon' = \tan\delta$ is referred to as the loss tangent or dissipation factor. Both σ and $\tan\delta$ are used in tables of dielectric characteristics of tissue. It is common to express the complex propagation constant as

$$\gamma = \alpha + j\beta \tag{3.33}$$

where α is the attenuation constant in nepers/meter and β is the phase constant in radians/meter.

The constants α and β are found by setting (3.32) equal to (3.33) and then squaring both sides, equating the real and imaginary parts, and solving the pair of simultaneous equations, with the result

$$\alpha = \frac{\omega\sqrt{\mu\epsilon'}}{\sqrt{2}}\{\sqrt{1 + (\frac{\sigma}{\omega\epsilon'})^2} - 1\}^{1/2} \tag{3.34}$$

$$\beta = \frac{\omega\sqrt{\mu\epsilon'}}{\sqrt{2}}\{\sqrt{1 + (\frac{\sigma}{\omega\epsilon'})^2} + 1\}^{1/2} \tag{3.35}$$

The wavelength λ in the lossy dielectric is then computed from the phase constant given by (3.35) as

$$\lambda = \frac{2\pi}{\beta} \tag{3.36}$$

The intrinsic wave impedance η is related to the ratio of the electric and magnetic fields and in lossy tissue is given by [28, pp. 333-334]

$$\eta = \sqrt{\frac{j\omega\mu}{\sigma + j\omega\epsilon'}} = \sqrt{\frac{\mu}{\epsilon'}}\frac{1}{\sqrt{1 - j\frac{\sigma}{\omega\epsilon'}}} \tag{3.37}$$

In free space, the electrical conductivity σ is equal to zero, and it follows from (3.37) that the free-space intrinsic wave impedance is

$$\eta_{\text{free space}} = \sqrt{\frac{\mu_o}{\epsilon_o}} = 120\pi \approx 377 \text{ ohms} \tag{3.38}$$

3.2.3 ELECTROMAGNETIC ENERGY FLOW

To demonstrate that an electromagnetic field can transfer electromagnetic energy into an electrically conducting volume (such as tissue), consider the following application of Maxwell's curl equations in the time domain. Starting with the curl of the instantaneous magnetic field as in (3.1), substitute (3.24) with the result

$$\nabla \times H(x, y, z, t) = \sigma E + \epsilon' \frac{\partial E}{\partial t} \tag{3.39}$$

Next, the curl of the instantaneous electric field from (3.2) is given by

$$\nabla \times E(x, y, z, t) = -\mu \frac{\partial H}{\partial t} \tag{3.40}$$

Fundamentally, it is desired to derive a power-flow expression for the electric and magnetic fields in tissue. The electric field has units of volts/meter and the magnetic field has units of amperes/meter. The product of volts/meter and amperes/meter produces watts/meter^2, which is power density. There are two possibilities for manipulating the electric and magnetic field expressions to produce a relation for power density, either a dot product or a cross product of the two field quantities. To produce a meaningful relation here, the cross product is the desired vector operation on the electric and magnetic field vectors. A vector identity will also be needed to convert to the desired field relation. Of the available general vector identities, the one that applies here is

$$\nabla \cdot (A \times B) = B \cdot (\nabla \times A) - A \cdot (\nabla \times B) \tag{3.41}$$

Based on the vector identity in (3.41), it observed that the desired mathematical manipulations are to dot the electric field into the curl of the magnetic field and to dot the magnetic field into the curl of the electric field. Once these manipulations are done, the desired cross product between E and H will appear. Thus, from (3.39), (3.40), and (3.41) and using $J \cdot E = E \cdot J$ it follows that

$$E \cdot (\nabla \times H) = J \cdot E + E \cdot \epsilon' \frac{\partial E}{\partial t} = J \cdot E + \frac{1}{2} \epsilon' \frac{\partial E^2}{\partial t} \tag{3.42}$$

and

$$H \cdot (\nabla \times E) = -H \cdot \mu \frac{\partial H}{\partial t} = -\frac{1}{2} \mu \frac{\partial H^2}{\partial t} \tag{3.43}$$

Simplifying and subtracting (3.43) from (3.42) yields

$$\boldsymbol{E}\cdot(\boldsymbol{\nabla}\times\boldsymbol{H}) - \boldsymbol{H}\cdot(\boldsymbol{\nabla}\times\boldsymbol{E}) = \boldsymbol{J}\cdot\boldsymbol{E} + \frac{1}{2}\epsilon'\frac{\partial E^2}{\partial t} + \frac{1}{2}\mu\frac{\partial H^2}{\partial t} \qquad (3.44)$$

or applying the vector identity (3.41) in (3.44) yields

$$\boldsymbol{\nabla}\cdot(\boldsymbol{E}\times\boldsymbol{H}) = -\boldsymbol{J}\cdot\boldsymbol{E} - \frac{1}{2}\epsilon'\frac{\partial E^2}{\partial t} - \frac{1}{2}\mu\frac{\partial H^2}{\partial t} \qquad (3.45)$$

and now integrating over a volume of tissue

$$\int_V \boldsymbol{\nabla}\cdot(\boldsymbol{E}\times\boldsymbol{H})dv = -\int_V \boldsymbol{J}\cdot\boldsymbol{E}dv - \frac{\partial}{\partial t}\int_V [\frac{1}{2}\epsilon'E^2 + \frac{1}{2}\mu H^2]dv \qquad (3.46)$$

The divergence theorem for any vector $\boldsymbol{A}$ for a closed surface S enclosing volume V is given by

$$\int_V \boldsymbol{\nabla}\cdot\boldsymbol{A}dv = \oint_S \boldsymbol{A}\cdot\hat{n}ds \qquad (3.47)$$

where dv is the elemental volume, and ds is the elemental area perpendicular to the outward unit normal $\hat{n}$. Thus, applying (3.47) to the left-hand side of (3.46) yields

$$\int_V \boldsymbol{\nabla}\cdot(\boldsymbol{E}\times\boldsymbol{H})dv = \oint_S (\boldsymbol{E}\times\boldsymbol{H})\cdot\hat{n}ds \qquad (3.48)$$

And finally, (3.46) reduces to

$$\oint_S (\boldsymbol{E}\times\boldsymbol{H})\cdot\hat{n}ds = -\int_V \boldsymbol{J}\cdot\boldsymbol{E}dv - \frac{\partial}{\partial t}\int_V [\frac{1}{2}\epsilon'E^2 + \frac{1}{2}\mu H^2]dv \qquad (3.49)$$

The term on the left side of (3.49) represents the net inward flux of the vector

$$\boldsymbol{P}(x,y,z,t) = \boldsymbol{E}(x,y,z,t)\times\boldsymbol{H}(x,y,z,t) \qquad (3.50)$$

In (3.49) and (3.50), the vector quantity $\boldsymbol{P}(x,y,z,t) = \boldsymbol{E}\times\boldsymbol{H}$ is known as the instantaneous Poynting vector (due to J.H. Poynting [40]) and can be interpreted as the instantaneous power flow density in watts/meter^2 at each point (x,y,z) on the surface S, or equivalently the power delivered to the volume V by external electromagnetic sources. Equation (3.49) is referred to as the integral form of Poynting's theorem and describes conservation of energy for electromagnetic waves. From (3.24) it follows that

$$\boldsymbol{J}\cdot\boldsymbol{E} = \sigma\boldsymbol{E}\cdot\boldsymbol{E} = \sigma|\boldsymbol{E}|^2 \qquad (3.51)$$

where the dot product of the real vector E with itself is just the magnitude of E squared. The first integral term on the right side of (3.49) involving $J{\cdot}E = \sigma|E|^2$ represents the instantaneous ohmic losses within the volume enclosed by the surface S. The integral in the second term on the right side of (3.49) is the total energy (with units of watt-seconds = joules), due to electric and magnetic fields, stored within the volume V.

To understand the electromagnetic field energy transfer and heating of conductive media such as tissue, it is necessary to determine the time-average power delivered to and dissipated within the tissue volume. The time-average Poynting vector or time-average power density given by (3.50) is computed over one period T as

$$P_{ave}(x, y, z) = \frac{1}{T} \int_0^T E(x, y, z, t){\times}H(x, y, z, t)dt \tag{3.52}$$

where the period is equal to

$$T = 2\pi/\omega \tag{3.53}$$

Substituting the exponential forms (3.7) and (3.8) into (3.52) yields

$$P_{ave}(x, y, z) = \frac{1}{T} \int_0^T \left[\text{Re}[E(x, y, z)e^{j\omega t}]{\times}\text{Re}[H(x, y, z)e^{j\omega t}] \right] dt \tag{3.54}$$

Making use of the relation

$$\text{Re}[AB] = \frac{1}{2}[AB + A^*B^*] \tag{3.55}$$

it follows that

$$\text{Re}[E(x, y, z)e^{j\omega t}]{\times}\text{Re}[H(x, y, z)e^{j\omega t}]$$
$$= \frac{1}{2}[Ee^{j\omega t} + E^*e^{-j\omega t}]{\times}\frac{1}{2}[He^{j\omega t} + H^*e^{-j\omega t}] \tag{3.56}$$

Expanding (3.56) and using the relation

$$\text{Re}[E{\times}H^*] = \frac{1}{2}[E{\times}H^* + E^*{\times}H] \tag{3.57}$$

it can be shown that (3.54) reduces to

$$P_{ave} = \frac{1}{T} \int_0^{\frac{2\pi}{\omega}} \left[\frac{1}{2}\text{Re}[E{\times}H^*] + \frac{1}{4}E{\times}He^{j2\omega t} + \frac{1}{4}E^*{\times}H^*e^{-j2\omega t} \right] dt \tag{3.58}$$

It can be shown, by integration, that (3.58) reduces to (see Problem 3.14)

$$P_{ave}(x, y, z) = \frac{1}{2}\text{Re}[E \times H^*] \tag{3.59}$$

which is the desired expression for the time-average power density of the electromagnetic field, or the time-average Poynting's vector, which has units of (W/m^2).

A plane wave can be used to approximate the field for hyperthermia treatments [3, pp. 243-249]. A plane wave is an oversimplification, but it provides a tractable example. As an example, for the case of a plane wave propagating in a conducting medium, the magnetic field is given by

$$H(x, y, z) = \hat{n} \times E(x, y, z)/\eta \tag{3.60}$$

and thus, substituting (3.60) in (3.59) it follows from a vector identity that the time-average Poynting vector for a plane wave is equal to

$$P_{ave}(x, y, z) = \hat{n}\frac{1}{2}|E|^2/\eta \tag{3.61}$$

where $\hat{n}$ is the direction of propagation.

Next, it is desired to quantify the average power deposition in tissue, because this quantity is needed to determine the absorption of electromagnetic energy in tissue. Referring to (3.49), for the first term on the right-hand side the time-average power dissipation per unit volume is computed over one period T as

$$P_d(x, y, z) = \frac{1}{T}\int_0^T J(x, y, z, t) \cdot E(x, y, z, t)dt \tag{3.62}$$

Substituting (3.7) and (3.9) into (3.62) yields

$$P_d(x, y, z) = \frac{1}{T}\int_0^T \left[\text{Re}[J(x, y, z)e^{j\omega t}] \cdot \text{Re}[E(x, y, z)e^{j\omega t}]\right] dt \tag{3.63}$$

Making use of the relation given in (3.55), it follows that

$$\text{Re}[J(x, y, z)e^{j\omega t}] \cdot \text{Re}[E(x, y, z)e^{j\omega t}]$$
$$= \frac{1}{2}[Je^{j\omega t} + J^*e^{-j\omega t}] \cdot \frac{1}{2}[Ee^{j\omega t} + E^*e^{-j\omega t}] \tag{3.64}$$

Using (3.64) in (3.63) and observing that the current density J and electric field E are in phasor form, it follows that

$$J \cdot E^* = \sigma E \cdot E^* = \sigma|E|^2 \tag{3.65}$$

$$\boldsymbol{J}^* \cdot \boldsymbol{E} = \sigma \boldsymbol{E}^* \cdot \boldsymbol{E} = \sigma |\boldsymbol{E}|^2 \tag{3.66}$$

and it can be shown that

$$P_d = \frac{1}{T} \int_0^{T=\frac{2\pi}{\omega}} \left[\frac{1}{2}\sigma |\boldsymbol{E}|^2 + \frac{1}{4} \boldsymbol{J} \cdot \boldsymbol{E} e^{j2\omega t} + \frac{1}{4} \boldsymbol{J}^* \cdot \boldsymbol{E}^* e^{-j2\omega t} \right] dt \tag{3.67}$$

It also can be shown that (3.67) reduces to

$$P_d(x, y, z) = P_{\text{ave dissipated}} = \frac{1}{2}\sigma |\boldsymbol{E}(x, y, z)|^2 \tag{3.68}$$

which is the desired expression for the time-average power dissipation per unit volume of tissue, which has units of (W/m^3).

The specific absorption rate (SAR) [1, p. 252], [2, p. 9], [3, p. 217] is the power dissipated or absorbed per unit mass (W/kg) of the medium (tissue), so from (3.68) it follows that

$$\text{SAR}(x, y, z) = \frac{P_d(x, y, z)}{\rho} = \frac{\sigma}{2\rho} |\boldsymbol{E}(x, y, z)|^2 \tag{3.69}$$

where ρ is the density of the medium in kg/m^3. To achieve therapeutic heating of tumors, the local SAR must be in the range of about 50 to 350 W/kg [2, pp. 20-21].

3.2.4 RF AND MICROWAVE PROPERTIES OF TISSUE

The microwave properties of tissues have been measured over a wide range of frequencies as described by a number of authors [4, 7-18]. The two primary frequencies of interest for the adaptive phased array thermotherapy system examples discussed in this book are in the vicinity of 100 and 900 MHz. Table 3.3 lists the measured dielectric constant, measured ionic conductivity, and calculated microwave attenuation (from (3.34) and (3.77)) of tissues including bladder, blood, bone, brain, normal fatty breast, colon, eyes, fat, kidney, liver, inflated lung, muscle, ovary, skin, and uterus, typically at normal body temperature; that is, 37°C. Also listed for comparison are the parameters for water and physiological saline (0.9%) at room temperature. Table 3.4 lists the electrical parameters for cancerous tissues including bladder, breast, colon, kidney, liver, and lung. These cancerous tissues have dielectric properties similar to muscle and should be readily heated by RF and microwave energy. In comparing the normal tissues and cancerous tissues in Tables 3.3 and 3.4, respectively, it is observed that normal fatty breast tissue and breast cancer tumors have significantly different values for dielectric constant, electrical conductivity, and attenuation. Refering to

Table 3.3

Estimated Mean Dielectric Parameters and Calculated Plane Wave Microwave Attenuation for Normal Body Tissues at 100 and 900 MHz Based on Published Measured Data

Tissue Type	Tissue Electrical Parameters at RF and Microwave Frequencies					
	Dielectric Constant		Conductivity (S/m)		Attenuation (dB/cm)	
	100 MHz	900 MHz	100 MHz	900 MHz	100 MHz	900 MHz
Bladder[g]	22.6	18.9	0.29	0.38	0.75	1.40
Blood[g]	76.8	66.4	1.2	1.5	1.59	3.00
Bone[g] (cortical)	15.1	12.4	0.06	0.14	0.24	0.65
Brain[d]	74.4	41.6	0.46	0.90	0.78	2.23
Breast[c,j] (fatty)	20.5	12.5	0.11	0.21	0.36	0.96
Colon[j]	52	48.5	0.62	0.93	1.08	2.15
Eye[d]	48	36	0.35	0.8	0.79	2.13
Fat[g]	6.1	5.5	0.04	0.05	0.24	0.35
Kidney[j]	72	61	0.77	1.20	1.18	2.47
Liver[j]	62	51.1	0.49	0.95	0.87	2.14
Lung[g] (inflated)	31.6	22	0.3	0.46	0.72	1.57
Muscle[g]	66	55	0.7	0.94	1.1	2.1
Ovary[g]	87.2	50.5	0.75	1.3	1.1	2.9
Prostate[g]	75.6	60.6	0.91	1.2	1.32	2.48
Skin[d]	65	44	0.78	1.1	1.22	2.64
Stomach[g]	77.9	65.1	0.90	1.2	1.3	2.39
Uterus[g]	80	61.1	0.94	1.27	1.33	2.51
Water, 25°C	78.3	78.1	0.002	0.17	0.004	0.32
Saline 0.9%	75.3	75.1	1.6	1.7	1.92	3.13

c=Chaudhary [12], d=Duck [4], j=Joines [13], g=Gabriel [16].

Table 3.1, normal breast tissue (containing fatty (adipose) tissue, glandular, and connective tissue) tends to be lower in water content (30% to 70%) when compared to breast cancer (66% to 79%), and it follows that the electrical conductivity of normal fatty breast tissue would be lower than that of breast cancer. For example, from Tables 3.3 and 3.4, at about 900 MHz, the electrical conductivity of normal breast is 0.21 S/m compared to 1.03 S/m for breast cancer.

Since the specific absorption rate (SAR) given by (3.69) is proportional to the electrical conductivity, for a given value of electric field it follows that breast cancer should heat significantly faster than normal fatty breast

Table 3.4
Estimated Mean Dielectric Parameters and Calculated Plane Wave Microwave
Attenuation for Cancerous Tumor Tissues at 100 and 900 MHz Based on Published
Measured Data (23°C to 25°C)

Cancerous Tumor Type	Electrical Parameters for Cancer at RF and Microwave Frequencies					
	Dielectric Constant		Conductivity (S/m)		Attenuation (dB/cm)	
	100 MHz	900 MHz	100 MHz	900 MHz	100 MHz	900 MHz
[j]Bladder CA	63	55	0.92	1.54	1.4	3.3
Breast CA	69[j]	58.6[c,j]	0.78[j]	1.03[c,j]	1.2	2.2
[j]Colon CA	65	56	0.74	1.08	1.2	2.2
[j]Kidney CA	68	57.6	0.75	1.20	1.2	2.5
[j]Liver CA	66	57	0.66	1.1	1.1	2.3
[g]Lung CA	69	54	0.82	1.24	1.3	2.3
[s]Water, 25°C	78.3	78.1	0.002	0.17	0.004	0.32
[s]Saline 0.9%	75.3	75.1	1.6	1.7	1.92	3.13

g=Gabriel [16], *j*=Joines [13], *s*=Stogryn [19].

tissue. The hypothesis that adaptively focused microwave energy can provide preferential heating of breast cancer compared to normal breast tissue is explored in Chapters 8 and 9. Lazebnik [18] has recently compared the dielectric parameters for normal, benign, and malignant breast tissues over the 0.5- to 20-GHz band, and the complex permittivity was characterized by a one-pole Cole-Cole model – further details of this study are reviewed in Chapter 8. It should be noted that the dielectric constant, electrical (ionic) conductivity, microwave phase propagation constant, and microwave attenuation are temperature-dependent [4, p. 173]. A number of phantom materials have been formulated to approximate the electrical characteristics of tissues, from low-water/low-ion content phantom simulating fat to high-water/high-ion content phantom simulating muscle [2, pp. 9-13], [41-44].

3.2.5 RAY-TRACING ANALYSIS TECHNIQUE

It is convenient to have a simple equation for computing the wave propagation between any two points in the near field of an isolated transmitting antenna in conducting media. Consider a time-harmonic source radiating a spherical wave into an infinite homogeneous conducting medium. For an isotropic radiator, and suppressing the $e^{j\omega t}$ time dependence where $\omega = 2\pi f$, the electric field as a function of range r can be expressed in phasor notation as a spherical wave

$$E(r) = E_o \frac{e^{-\gamma r}}{r} \qquad (3.70)$$

where E_o is a constant and γ is the propagation constant given by (3.32). In some cases where the $1/r$ dependence can be ignored, a plane wavefront approximation to (3.70) can be used. For example, a plane wavefront propagating along the x-axis can be expressed as

$$E(x) = E_o e^{-\gamma x} \qquad (3.71)$$

Equations (3.70) or (3.71) can be used in a simplified ray-tracing analysis to compute the approximate electric field, generated by one or more transmitting antennas, at any point in a homogeneous tissue phantom. Letting the electric field at the rectangular coordinates (x, y, z) due to the nth transmitting antenna in an array of N transmitting antennas be denoted as $E_n(x, y, z)$, the total electric field of the transmitting array can be expressed as the following summation,

$$E_{\text{total}}(x, y, z) = \sum_{n=1}^{N} E_n(x, y, z) \qquad (3.72)$$

For a transmitting source at the origin, the amplitude of the electric field at range r_1 is given by

$$|E(r_1)| = E_o \frac{e^{-\alpha r_1}}{r_1} \qquad (3.73)$$

and at range r_2 by

$$|E(r_2)| = E_o \frac{e^{-\alpha r_2}}{r_2} \qquad (3.74)$$

The total propagation loss between ranges r_1 and r_2 is found by taking the ratio of (3.74) and (3.73), or

$$\frac{|E(r_2)|}{|E(r_1)|} = \frac{r_1}{r_2} e^{-\alpha(r_2 - r_1)} \qquad (3.75)$$

The field attenuation A_α in decibels from range r_1 to range r_2 due to the lossy dielectric is simply

$$A_\alpha = 20 \log_{10}(e^{-\alpha(r_2 - r_1)}) \qquad (3.76)$$

Letting $d = r_2 - r_1$, the above equation can be written in a convenient form as

$$A_\alpha = 20 \log_{10}(e^{-\alpha d}) \qquad (3.77)$$

which is the relative field attenuation as a function of distance d into tissue or transmission loss of a plane wave in decibels per meter. The attenuation constant α has units of nepers per unit length or typically in the literature of nepers per meter (Np/m), where neper is dimensionless. If d is one meter and $\alpha = 1$ Np/m, then the field strength in decibels is $20 \log_{10}(e^{-1}) = 20 \log_{10} |0.368| = -8.686$ dB. Thus, the attenuation in decibels per meter is 8.686 times the attenuation constant in nepers per meter. If the attenuation constant is expressed in nepers/cm, as appropriate for hyperthermia treatment of cancer, then to convert to attenuation in decibels the multiplying factor remains 8.686. Similarly, the $1/r$ attenuation loss A_r in decibels is

$$A_r = 20 \log_{10} \frac{r_1}{r_2} \tag{3.78}$$

An example of coherent opposing plane waves irradiating lossy tissue is shown in Figure 3.2. The two electric fields have different phase shifts ϕ_1 and

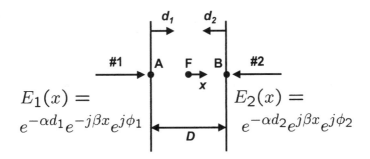

Figure 3.2 Diagram for coherent opposing waves irradiating lossy tissue of thickness D.

ϕ_2, and can be expressed in phasor notation as $E_1(x) = E_o e^{-\alpha d_1} e^{-j\beta x} e^{j\phi_1}$, and $E_2(x) = E_o e^{-\alpha d_2} e^{j\beta x} e^{j\phi_2}$. The total electric field is then given by $E_T(x) = E_1(x) + E_2(x)$. The total electric field and specific absorption rate can be determined for various tissues and can be used in simulations of adaptive nulling and adaptive focusing as given in the Problem Set (Problem 3.13) in this chapter.

It should be understood that the simplified ray-tracing approach described in this section for plane waves or spherical waves provides only an approximate model for electromagnetic analysis of an array applicator for heating tissue. A very accurate approach for modeling arbitrary heterogeneous tissue bodies is the finite-difference time-domain method. However, for homogeneous tissues a method of moments approach as described in the next section can be used.

3.3 METHOD OF MOMENTS FORMULATION

The method of moments [45-50] is a numerical technique based on a solution of Maxwell's equations that is used to compute the electromagnetic fields of antennas and scattering objects such as in the application of hyperthermia [2, pp. 14-15]. This section describes a method of moments formulation to compute the probe-received voltages in the cross-correlation expression given by (2.3) due to the transmitting hyperthermia phased array antenna in an infinite homogeneous conducting medium. This formulation is applied to an adaptive phased array in Chapter 5. The medium is described by the parameters μ and ϵ that were discussed in Section 3.2.4. The formulation given here is analogous to that developed under array-receiving conditions for an adaptive phased array radar [21, pp. 111-131]. The theory presented below is given in the context of an adaptive transmit phased array.

Referring to Figure 3.3, assume that each transmit array antenna element is fed with a generator having a known impedance Z_L, and assume that the generators are coherent with respect to each other. The generators are weighted by a complex weight vector denoted as $w = (w_1, w_2, \cdots, w_n, \cdots, w_N)^T$, which is interpreted here as a normalized voltage vector. The nth weight is expressed in terms of an amplitude a_n and phase ϕ_n as

$$w_n = a_n e^{j\phi_n} \qquad (3.79)$$

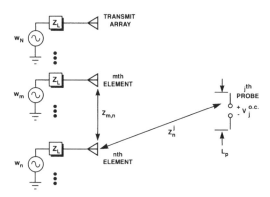

Figure 3.3 Transmit phased array with coherent generators and receive antenna probe. Open-circuit mutual impedance between array elements is denoted $Z_{m,n}$. The open-circuit voltage at the probe is computed from the array terminal currents and from Z_n^j, the open-circuit mutual impedance between the nth array element and the jth probe antenna.

Let $v_{n,j}^{o.c.}$ represent the open-circuit voltage at the jth probe due to the nth transmit-array element. Here, the jth probe can denote either the focal point calibration probe or one of the auxiliary probes used to null the field at healthy tissues (see Figure 1.10). The number of auxiliary probes is denoted by N_{aux}. The receive probe is assumed to be terminated in an impedance Z_r. Next, let Z denote the open-circuit mutual impedance matrix (with dimensions $N \times N$ for the N-element hyperthermia phased array). The open-circuit mutual impedance between array elements m and n is denoted $Z_{m,n}$. It is assumed that multiple interaction between the hyperthermia array and the probe sensor can be neglected. Thus, as shown in this section (see (3.101)) the hyperthermia array terminal current vector i can be computed in terms of the transmit weights w as

$$i = [Z + Z_L I]^{-1} w \tag{3.80}$$

where I is the identity matrix.

Now, assume that the hyperthermia transmit array is composed of thin-wire antennas such as dipoles, and that the receive probe is also a dipole antenna. Next, let Z_n^j be the open-circuit mutual impedance between the jth probe and the nth array element. The open-circuit mutual impedances can be computed for thin-wire antennas by assuming sinusoidal current basis functions as described by Richmond [45-48]. The induced open-circuit voltage $v_{n,j}^{o.c.}$ at the jth receive probe, due to the nth array element transmit current i_n, can then be expressed as

$$v_{n,j}^{o.c.} = Z_n^j i_n \tag{3.81}$$

In matrix form, the induced open-circuit probe-voltage matrix $v_{probe}^{o.c.}$ is

$$v_{probe}^{o.c.} = Z_{probe,array} i \tag{3.82}$$

or substituting (3.80) in (3.82) yields

$$v_{probe}^{o.c.} = Z_{probe,array} [Z + Z_L I]^{-1} w \tag{3.83}$$

where $Z_{probe,array}$ is a rectangular matrix of order $N_{aux} \times N$ for the open-circuit mutual impedance between the probe array and the hyperthermia array. Note that the jth row of the matrix $Z_{probe,array}$ is written as $(Z_1^j, Z_2^j, \cdots, Z_N^j)$, where $j = 1, 2, \cdots, N_{aux}$. The receive voltage matrix is then computed by the receiving circuit equivalence theorem for an antenna [39, pp. 294-296] The receive antenna equivalent circuit is depicted in Figure 3.4, where it is readily determined that

$$v_{probe}^{rec} = v_{probe}^{o.c.} \frac{Z_r}{Z_{in} + Z_r} \tag{3.84}$$

where Z_{in} is the input impedance of the probe. It should be noted that the v_{probe}^{rec} matrix is a column vector of length N_{aux} and v_j^{rec} is the jth element of the matrix. From (3.84), it follows that the probe-receive current matrix is given by

$$i_{probe}^{rec} = v_{probe}^{o.c.} \frac{1}{Z_{in} + Z_r} \tag{3.85}$$

The jth element of the column vector i_{probe}^{rec} is denoted $i_j^{rec}, j = 1, 2, \cdots, N_{aux}$. Finally, the power received by the jth probe is

$$p_j^{rec} = \frac{1}{2} \mathrm{Re}(v_j^{rec} i_j^{rec*}) \tag{3.86}$$

where Re means real part. Substituting (3.84) and (3.85) into (3.86) yields

$$p_j^{rec} = \frac{1}{2} |v_j^{o.c.}|^2 \frac{\mathrm{Re}(Z_r)}{|Z_{in} + Z_r|^2} \tag{3.87}$$

The total interference power received by the auxiliary probe array is given by

$$p^{rec} = \sum_{j=1}^{N_{aux}} p_j^{rec} \tag{3.88}$$

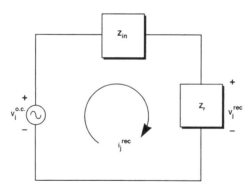

Figure 3.4 Equivalent circuit model for receive antenna probe.

The incident electric field E is related to the open-circuit voltage $v^{o.c.}$ by the effective height h of the probe antenna [38] as

$$v^{o.c.} = hE \tag{3.89}$$

If the length L_p of the probe antenna is approximately 0.1λ or less, the current distribution is triangular and the effective height is $h = 0.5L_p$ [38]. Thus, for a short-dipole probe the open-circuit voltage can be expressed as

$$v^{o.c.} = \frac{L_p}{2} E \tag{3.90}$$

It then follows from (3.90) that the E field for a short-dipole probe at position (x, y, z) is given by

$$E(x, y, z) = \frac{2v^{o.c.}(x, y, z)}{L_p} \tag{3.91}$$

Finally, the quiescent and adapted E-field radiation patterns are computed using the quiescent and adapted weight vectors w_q and w_a, respectively, from (2.4) and (2.5) and applied to (3.83) and (3.91).

In the method of moments formulation used here, a boundary condition is enforced such that the total tangential component of the electric field must be zero at each of the dipole elements in the hyperthermia array. The moment-method expansion and testing functions are assumed here to be sinusoidal, which is a good approximation to the current distribution on a dipole antenna. The open-circuit mutual impedances in (3.83) between thin-wire dipoles in a homogeneous conducting medium are computed based on subroutines from a well-known moment-method computer code [45-48]. In evaluating Z_n^j for the jth auxiliary probe, double precision computations are used.

Mathematically, in the electric-field integral equation form of the method of moments [49, 50, pp. 429-438], the boundary condition that must be satisfied is that the tangential electric field E_m must be zero at the mth dipole element. That is,

$$E_m = E_m^s + E_m^i = 0, \quad m = 1, 2, \ldots, N \tag{3.92}$$

where E_m^i is the incident field at dipole m, E_m^s is the scattered field at dipole m, and N is the number of array elements.

The scattered field can be expressed as the superposition of the fields radiated by the dipole currents $i_n(z)$, $n = 1, 2, \ldots, N$. For the case where the dipoles are z-directed, the currents can be assumed to be of the sinusoidal form

$$i_n(z) = i_n \frac{\sin \gamma(L/2 - |z|)}{\sin \gamma L/2} \tag{3.93}$$

where γ is the complex propagation constant given by (3.33), L is the dipole length, and i_n is the complex terminal current of the nth dipole. The normalized sinusoidal function in (3.93) is referred to here as the basis function $b_n(z)$. The dipole terminal current i_n is a complex number that is to be determined.

The z-component of the scattered electric field is written as

$$E_m^s(z) = \sum_{n=1}^{N} i_n \int_{-\frac{L}{2}}^{\frac{L}{2}} K(\boldsymbol{r}_m, \boldsymbol{r}_n') b_n(z') dz' \tag{3.94}$$

where $K(\boldsymbol{r}_m, \boldsymbol{r}_n')$ is the kernal function (or Green's function), $\boldsymbol{r}_m$ is the observation position vector on dipole m, and $\boldsymbol{r}_n'$ is the source position vector on dipole n.

Substituting (3.94) in (3.92) yields

$$E_m^i(z) = -\sum_{n=1}^{N} i_n \int_{-\frac{L}{2}}^{\frac{L}{2}} K(\boldsymbol{r}_m, \boldsymbol{r}_n') b_n(z') dz' \tag{3.95}$$

where $m = 1, 2, \ldots, N$.

A Galerkin's formulation is obtained by the integral weighting of (3.95) with testing functions $t_m(z)$ that have the same sinusoidal distribution as the basis functions $b_n(z)$. Equation (3.95) now becomes

$$\int_{-\frac{L}{2}}^{\frac{L}{2}} t_m(z) E_m^i(z) dz = -\sum_{n=1}^{N} i_n \int_{-\frac{L}{2}}^{\frac{L}{2}} \int t_m(z) K(\boldsymbol{r}_m, \boldsymbol{r}_n') b_n(z') dz' dz \tag{3.96}$$

where $m = 1, 2, \ldots, N$, or

$$V_m = \sum_{n=1}^{N} i_n Z_{mn} \quad m = 1, 2, \ldots, N \tag{3.97}$$

where

$$V_m = \int_{-\frac{L}{2}}^{\frac{L}{2}} t_m(z) E_m^i(z) dz \tag{3.98}$$

is the voltage excitation (which can be expressed as a column vector $\boldsymbol{V}$) and

$$Z_{mn} = -\int_{-\frac{L}{2}}^{\frac{L}{2}} \int t_m(z) K(\boldsymbol{r}_m, \boldsymbol{r}_n') b_n(z') dz' dz \tag{3.99}$$

is the open-circuit mutual impedance between elements m and n, which can be expressed as the matrix $\boldsymbol{Z}$. The open-circuit mutual impedance matrix elements are computed using the approach given by Richmond and Geary [48]. For convenience, a delta gap model can be assumed for the incident electric field at the dipole feed point [50, p. 435]. Equation (3.97) can be expressed in terms of the applied voltage excitation vector ($\boldsymbol{V}$), the impedance matrix Z, the generator load impedance Z_L, and the unknown current vector (i) to give the matrix equation

$$V = [Z + Z_L I]\, i \tag{3.100}$$

The unknown current coefficients $i_n, n = 1, 2, \ldots, N$ are then found by matrix inversion from

$$i = [Z + Z_L I]^{-1}\, V \tag{3.101}$$

As mentioned in Chapters 1 and 2, the hyperthermia transmit phased array is calibrated (phased focused) initially using a short dipole at the focal point. To accomplish this numerically, having computed v^{rec}_{focus}, the receive array weight vector w will have its phase commands set equal to the conjugate of the corresponding phases in v^{rec}_{focus}. Transmit antenna radiation patterns are obtained by scanning (moving) a dipole probe with half-length l in the near-field and computing the receive probe-voltage response.

For an arbitrary transmit waveform, the received voltage matrix for the jth probe (denoted v^{rec}_j) is computed at K frequencies across the nulling bandwidth. Thus, $v^{rec}_j(f_1), v^{rec}_j(f_2), \cdots, v^{rec}_j(f_K)$ are needed. The impedance matrix is computed at K frequencies and is inverted K times. The probe covariance matrix elements are computed by evaluating (2.2) numerically, using Simpson's rule numerical integration. For the CW case, as normally used in hyperthermia, the covariance matrix elements are computed by (2.3). For multiple auxiliary probes, the covariance matrix is evaluated using (2.4). This method of moments formulation is applied to an adaptive phased array of dipole-radiating elements in Chapter 5. For analysis of electromagnetic fields in heterogeneous tissues, the finite-difference time-domain method can be used as described in the next section.

3.4 FINITE-DIFFERENCE TIME-DOMAIN METHOD

The finite-diference time-domain method has been used extensively in the literature for general analysis of electromagnetic problems [51-55] including hyperthermia treatment of cancer [56, 57]. This method is well suited for analysis of electromagnetic fields in heterogeneous tissues. The finite-difference time-domain method can solve Maxwell's equations for arbitrary media by dividing the media into cubic cells in three dimensions. In the case of electromagnetic hyperthermia treatment of human tissues, the tissue volume of interest is divided into a volumetric collection of contiguous cubic cells, with each cubic cell representing the electrical properties of the tissue. The size of the cubic cells is selected to provide the required accuracy in modeling the tissue geometry.

As described in Section 3.2, in the time domain, Maxwell's curl equations can be expressed as

$$\nabla \times \boldsymbol{E} = -\mu \frac{\partial \boldsymbol{H}}{\partial t} \tag{3.102}$$

$$\nabla \times \boldsymbol{H} = \boldsymbol{J} + \epsilon' \frac{\partial \boldsymbol{E}}{\partial t} \tag{3.103}$$

where in rectangular coordinates

$$\boldsymbol{E} = E_x \hat{\boldsymbol{x}} + E_y \hat{\boldsymbol{y}} + E_z \hat{\boldsymbol{z}} \tag{3.104}$$

$$\boldsymbol{H} = H_x \hat{\boldsymbol{x}} + H_y \hat{\boldsymbol{y}} + H_z \hat{\boldsymbol{z}} \tag{3.105}$$

$$\boldsymbol{J} = J_x \hat{\boldsymbol{x}} + J_y \hat{\boldsymbol{y}} + J_z \hat{\boldsymbol{z}} \tag{3.106}$$

where $\boldsymbol{E}$ is the electric field, $\boldsymbol{H}$ is the magnetic field, and $\boldsymbol{J}$ is the electric current density. In (3.102) and (3.103), the curl operations are expressed as

$$\nabla \times \boldsymbol{E} = \begin{vmatrix} \hat{\boldsymbol{x}} & \hat{\boldsymbol{y}} & \hat{\boldsymbol{z}} \\ \frac{\partial}{\partial x} & \frac{\partial}{\partial y} & \frac{\partial}{\partial z} \\ E_x & E_y & E_z \end{vmatrix} \tag{3.107}$$

$$\nabla \times \boldsymbol{H} = \begin{vmatrix} \hat{\boldsymbol{x}} & \hat{\boldsymbol{y}} & \hat{\boldsymbol{z}} \\ \frac{\partial}{\partial x} & \frac{\partial}{\partial y} & \frac{\partial}{\partial z} \\ H_x & H_y & H_z \end{vmatrix} \tag{3.108}$$

Thus, in rectangular coordinates, Maxwell's curl equations ((3.102) and (3.103)) become:

$$\frac{\partial H_x}{\partial t} = \frac{1}{\mu} \left(\frac{\partial E_y}{\partial z} - \frac{\partial E_z}{\partial y} \right) \tag{3.109}$$

$$\frac{\partial H_y}{\partial t} = \frac{1}{\mu} \left(\frac{\partial E_z}{\partial x} - \frac{\partial E_x}{\partial z} \right) \tag{3.110}$$

$$\frac{\partial H_z}{\partial t} = \frac{1}{\mu} \left(\frac{\partial E_x}{\partial y} - \frac{\partial E_y}{\partial x} \right) \tag{3.111}$$

$$\frac{\partial E_x}{\partial t} = \frac{1}{\epsilon'} \left(\frac{\partial H_z}{\partial y} - \frac{\partial H_y}{\partial z} - \sigma E_x \right) \tag{3.112}$$

$$\frac{\partial E_y}{\partial t} = \frac{1}{\epsilon'} \left(\frac{\partial H_x}{\partial z} - \frac{\partial H_z}{\partial x} - \sigma E_y \right) \tag{3.113}$$

$$\frac{\partial E_z}{\partial t} = \frac{1}{\epsilon'} \left(\frac{\partial H_y}{\partial x} - \frac{\partial H_x}{\partial y} - \sigma E_z \right) \tag{3.114}$$

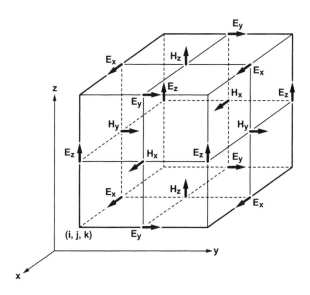

Figure 3.5 Positions of the electric and magnetic field components about a unit cell of the Yee lattice. (Redrawn from Taflove and Brodwin [52].)

Referring to Figure 3.5, following Yee's and Taflove's finite-difference time-domain formulations [51, 52], assume that points in space are arranged in a regular grid and are spaced in increments of $\Delta x = \Delta y = \Delta z = d$. Let the (i, j, k)th point in space with coordinates (x, y, z) be given by

$$(x, y, z) = (id, jd, kd) \tag{3.115}$$

Furthermore, let the time increment be given by Δt. Thus, at the nth time increment, any of the functions such as $E_x, E_y, E_z, H_x, H_y, H_z$ at the (i, j, k)th point in space can be expressed as

$$F(i, j, k, n) = F(id, jd, kd, n\Delta t) \tag{3.116}$$

The desired finite differences can be approximated as:

$$\frac{\partial F(i, j, k, n)}{\partial x} = \frac{F(id + d/2, jd, kd, n\Delta t) - F(id - d/2, jd, kd, n\Delta t)}{d} \tag{3.117}$$

$$\frac{\partial F(i, j, k, n)}{\partial y} = \frac{F(id, jd + d/2, kd, n\Delta t) - F(id, jd - d/2, kd, n\Delta t)}{d} \tag{3.118}$$

$$\frac{\partial F(i, j, k, n)}{\partial z} = \frac{F(id, jd, kd + d/2, n\Delta t) - F(id, jd, kd - d/2, n\Delta t)}{d} \tag{3.119}$$

and

$$\frac{\partial F(i,j,k,n)}{\partial t} = \frac{F(id,jd,kd,n\Delta t + \Delta t/2) - F(id,jd,kd,n\Delta t - \Delta t/2)}{\Delta t}$$

(3.120)

Small values of d and Δt are chosen to achieve accuracy and stability of the fields. For example, the lattice spacing must be small in terms of a wavelength to avoid significant field variation over one space interval. The cubic approximation to a smooth scatterer must have sufficiently fine detail for accurate field computation. Truncation of the lattice must be done in such a way that reflections are not introduced at the lattice boundary. To avoid lattice truncation reflections, a soft lattice truncation method (absorbing boundary condition) as described by Taflove is used [52].

For a phased array antenna, the source modeling can be of the resistive source type described by Reuter [57] as used in the simulations presented in Chapters 7 and 8. Since Maxwell's equations are linear, the field distribution of an antenna array with all elements radiating in a linear medium can be represented equivalently as the linear superposition of the field distributions due to each element radiating individually, as in (3.72).

3.5 SUMMARY

This chapter has provided a background for computation of the electromagnetic field in homogeneous and heterogeneous tissues. The theory for electromagnetic power flow and deposition in tissue has been reviewed. A simplified ray-tracing approach was described that can be used for approximate analysis of electromagnetic wave propagation in homogeneous tissue. A method of moments formulation was described that can be used for analyzing an electromagnetic phased array irradiating homogeneous tissue. The finite-difference time-domain method can be used in analyzing arbitrary heterogeneous tissues irradiated by electromagnetic waves. The next chapter shows how to simulate the thermal distribution induced by the electromagnetic fields that irradiate tissue.

3.6 PROBLEM SET

3.1 Refer to the diagram showing opposing plane waves in Figure 3.2. Assume that the irradiated tissue is lossy and has thickness D, and assume that the two plane waves are coherent and in phase. For convenience, assume that ϕ_1 and ϕ_2 are both zero. Then, let $E_1(x) = E_o e^{-\alpha d_1} e^{-j\beta x}$ and $E_2(x) = E_o e^{-\alpha d_2} e^{j\beta x}$. **a)** Show that the total electric field $E_T(x)$ in the lossy dielectric between points A and B is given by $E_T(x) = 2E_o e^{-\alpha \frac{D}{2}} \cosh(\gamma x)$ and thus it follows that $\text{SAR}(x) = \frac{\sigma}{2\rho} 4 |E_o|^2 e^{-\alpha D} \cosh^2(\gamma x)$. Hint: use $d_1 = D/2 + x$ and $d_2 = D/2 - x$, $\gamma = \alpha + j\beta$, and $e^Z + e^{-Z} = 2\cosh Z$, where Z is a complex number. **b)** Show that for the lossless case in free space, the total electric field $E_T(x)$ of part (a) reduces to (2.33).

3.2 From Table 3.3, fatty breast tissue in the vicinity of the microwave frequency 915 MHz has a dielectric constant of 12.5 and an electrical conductivity of 0.21 S/m. **a)** Using (3.34) and (3.35), show that the attenuation constant α is 0.11 nepers/cm and the phase constant β is 0.69 radians/cm. **b)** Using (3.36), show that the 915-MHz wavelength in fatty breast tissue is 9.1 cm. **c)** The free-space wavelength can be computed from $\lambda_o = c/f$, where c is the speed of light. Show that the free-space wavelength at 915 MHz is 32.8 cm. Thus, from part (b) the wavelength in breast tissue is reduced by a factor of 3.6 compared to the free-space wavelength. Equivalently, the wavelength in the breast tissue compared to the wavelength in free space is reduced by approximately the square root of the dielectric constant of breast tissue, that is, $\sqrt{\epsilon'_r} = \sqrt{12.5} = 3.54$. **d)** Using (3.37), show that the intrinsic wave impedance in fatty breast tissue at 915 MHz is $\eta = 102.5 + j16.5$ ohms. In contrast the intrinsic wave impedance in free space is $\eta_o = 120\pi = 377$ ohms. **e)** Using the results in part (a) for $\alpha = 0.11$ nepers/cm and $\beta = 0.69$ radians/cm for breast tissue, graph $E_T(x)$ as determined in Problem 3.1(a) for the cases where $D = 4.5$ cm, $D = 6$ cm, and $D = 8$ cm.

3.3 Repeat Problem 3.2 parts (a) to (e), but now with muscle tissue instead of breast tissue. From Table 3.3, muscle tissue in the vicinity of the microwave frequency 915 MHz has a dielectric constant of 50.0 and an electrical conductivity of 1.3 S/m. Answer: The attenuation constant is 0.33 nepers/cm, the phase constant is 1.40 radians/cm, the wavelength is 4.5 cm, and the intrinsic wave impedance is $48.9 + j11.8$ ohms. From the results of part (e) in this problem and Problem 3.2(e), note that relative to the central focused field, the surface fields for the case of muscle tissue are significantly stronger than for breast tissue.

3.4 The ray-tracing discussion surrounding (3.71) applies to a homogeneous medium. The ray-tracing analysis can be made more general to allow boundaries between different slabs of tissues by allowing reflection and transmission coefficients. Referring to Figure 3.1, if a plane wave is propagating from medium 1 to medium 2, at normal angle of incidence the voltage reflection coefficient is given by $\Gamma = (\eta_2 - \eta_1)/(\eta_2 + \eta_1)$, and the voltage transmission coefficient is given by $T = 2\eta_2/(\eta_2 + \eta_1)$, where η_1 and η_2 are the intrinsic impedances of media 1 and 2, respectively. Also, the relation between the transmission coefficient and reflection coefficient is $T = 1 + \Gamma$. These equations can be derived by the reader by enforcing the boundary condition that the total tangential electric and magnetic fields must be continous across the boundary between the two different media. If the incident electric field has amplitude E_o at the interface between the two different tissue media, the reflected field is ΓE_o and the transmitted field is TE_o. Using the values of intrinsic impedance for fatty breast tissue and muscle tissue obtained in Problems (3.2) and (3.3), respectively, compute the reflection and transmission coefficients for normal angle of incidence at 915 MHz, where medium 1 is normal fatty breast tissue and medium 2 is muscle tissue. Answer: $\Gamma = -0.35 + j0.03$, $T = 0.65 + j0.03$.

3.5 Using the discussion in Problem 3.4, consider a 915-MHz plane wave propagating through skin and reaching, at normal angle of incidence, the boundary between the layer of skin (medium 1) and normal fatty breast tissue (fat) (medium 2). Compute the reflection and transmission coefficients at the boundary between the skin layer and fat layer. Hint: From Table 3.3, skin has a dielectric constant of 44.0 and an electrical conductivity of 1.1 S/m at 915 MHz. From (3.37) the intrinsic wave impedance of skin is computed to be $\eta_{skin} = \eta_1 = 52.4 + j12.2$ ohms. At 915 MHz, breast fat has a dielectric constant of 12.5 and electrical conductivity 0.21 S/m, from which the intrinsic impedance is computed to be $\eta_{fat} = \eta_2 = 102.5 + j16.5$ ohms. Answer: $\Gamma = 0.318 - j0.031$, $T = 1.318 - j0.031$.

3.6 In Problem 3.5, it was observed that the transmission coefficient T at the skin/breast-fat interface has a magnitude of 1.32 at 915 MHz. Thus, it is implied that the transmitted electric-field intensity in medium 2 increases in magnitude by $20 \log_{10} 1.32 = 2.41$ dB compared to the incident field at the skin/fat interface. Quantify the plane-wave power density and specific absorption rate across the skin/fat interface. Answer: From the discussion following (3.61), the plane-wave power density in a tissue medium is expressed as $|P| = 0.5|E|^2/|\eta|$. Let $|E_i|$ be the magnitude of the electric-field incident on the tissue interface. Thus, on the skin side of the skin/fat interface, the incident power density is $P_{skin} = 0.5|E_i|^2/53.8 = 0.0093|E_i|^2$. The transmitted plane-wave power density on the breast-fat side of the skin/breast-fat interface is $P_{fat} = 0.5|1.32E_i|^2/103.8 = 0.00839|E_i|^2$. Taking the ratio of the plane-wave power density on the fat side to that on the skin side yields a loss in power density of $10 \log_{10}(0.00839/0.0093) = 0.5$ dB. Now compute the loss due to reflection coefficient that is given by the following equation,

Mismatch Loss $= 10 \log_{10}(1 - |\Gamma|^2) = 10 \log_{10}(1 - |0.318|^2) = 0.5$ dB. Thus, the loss in power density crossing the interface is due to reflection mismatch at the interface between the skin and breast-fat layers. To investigate this effect further, it is necessary to compute the specific absorption rate across the tissue boundary. First, it is necessary to compute the total field E_{skin} at the skin/breast-fat interface on the skin side, which is given by $E_{skin} = E_i + E_r = E_i + \Gamma E_i = (1 + \Gamma)E_i = (1.318 - j0.031)E_i$, which has a magnitude of $|E_{skin}| = 1.32|E_i|$. By (3.69) on the skin side, $\text{SAR}_{skin} = \frac{\sigma}{2\rho}|E_{skin}|^2$, where E_{skin} is the total field on the skin side of the skin/fat interface. Since $|E_{skin}| = 1.32|E_i|$, it follows that $\text{SAR}_{skin} = \frac{\sigma}{2\rho}|1.32E_i|^2$. Referring to Table 4.1, the density of skin is 1.0 g/cm^3 and the density of fat is 0.85 g/cm^3. On the skin side of the skin/fat tissue interface the SAR is $\text{SAR}_{skin} = \frac{1.1}{2 \times 1.0}|1.32E_i|^2 = 0.958|E_i|^2$. On the breast-fat side of the skin/breast-fat interface the SAR is $\text{SAR}_{fat} = \frac{0.21}{2 \times 0.85}|1.32E_i|^2 = 0.215|E_i|^2$. Although the electric-field amplitude has increased in the breast-fat layer, the SAR is lower in the fat layer by a factor of $0.958/0.215 = 4.45$ because the power deposition in breast fat is much lower than that in skin. This reduced SAR is primarily due to the lower electrical conductivity of breast fat.

3.7 Derive the electric-field form of the wave equation expressed by (3.31). Then verify that the propagation constant is given by (3.32). Hint: First take the curl of (3.19) and then use (3.25). Next, use the vector identity, $\nabla \times \nabla \times A = \nabla(\nabla \cdot A) - \nabla^2 A$. Then use the Maxwell divergence equation for the electric field in a source-free region $\nabla \cdot E = 0$.

3.8 Derive the magnetic-field form of the wave equation expressed by $\nabla^2 H - \gamma^2 H = 0$. Hint: First take the curl of (3.25) and then use (3.19). Next, use the vector identity, $\nabla \times \nabla \times A = \nabla(\nabla \cdot A) - \nabla^2 A$. Then use the Maxwell divergence equation for the magnetic field $\nabla \cdot H = 0$.

3.9 Prove that a plane wave propagating along the x-axis given by $E(x) = e^{-\gamma x}$ is a solution of the electric-field form of the wave equation expressed by (3.10). Hint: Express the wave equation in rectangular coordinates as $\nabla^2 E - \gamma^2 E = \partial^2 E / \partial x^2 - \gamma^2 E = 0$ and substitute the given plane-wave form.

3.10 Prove that a spherical wave propagating in the radial direction given by $E(r) = e^{-\gamma r}/r$ is a solution of the electric-field form of the wave equation expressed by (3.31). Hint: Express the wave equation in spherical coordinates as $\nabla^2 E - \gamma^2 E = \frac{1}{r}\frac{\partial}{\partial r}(r^2\frac{\partial E}{\partial r}) - \gamma^2 E = 0$ and substitute the given spherical wave form.

3.11 Derive the attenuation constant given by (3.34) and the phase constant given by (3.35) by using (3.32) and (3.33).

3.12 Derive the Poynting vector given by (3.50) and the time-average power dissipation per unit volume given by (3.68). Hint: Much of this derivation is given in the text. Start by taking the divergence of the cross product between the electric and magnetic fields, and apply a vector identity, that is, $\nabla \cdot (E \times H) = H \cdot (\nabla \times E) - E \cdot (\nabla \times H)$. Next, substitute Maxwell's curl equations and then integrate over a volume V. Finally, apply the divergence theorem for any vector field A as $\int_V \nabla \cdot A \, dv = \oint_S A \cdot \hat{n} \, dS$ where $\hat{n}$ is a unit normal vector to the surface S.

3.13 Assume dual-opposing plane waves in lossy tissue as in Figure 3.2 with plane waves numbers 1 and 2 having a phase shift of ϕ_1 and ϕ_2 degrees, respectively. Thus, the electric fields can be expressed as $E_1(x) = E_o e^{-\alpha d_1} e^{-j\beta x} e^{j\phi_1}$, and $E_2(x) = E_o e^{-\alpha d_2} e^{j\beta x} e^{j\phi_2}$. **a)** Determine the equations for the total electric field $(E_T(x))$ and specific absorption rate (SAR(x)) due to the two plane waves. **b)** Assume a 6-cm homogeneous layer of fatty breast tissue and use the values of attenuation constant ($\alpha = 0.11$ Np/cm and $\beta = 0.69$ radians/cm). Let $\phi_1 = 0°$, $\phi_2 = 180°$, and assume that a central focus at $x = 0$ is desired. Subsititute these phase values into the equation determined in part (a) for the SAR that will be used as a feedback signal for demonstrating adaptive phase focusing. Hint: Refer to Section 8.6. Use the standard gradient-search algorithm in Chapter 2 with (2.55) and (2.56) to show that the adaptively focused condition converges to $\phi_2 = 0°$. **c)** Using the same parameters as part (b), use the fast-acceleration gradient-search algorithm to show that the adaptively focused condition converges to $\phi_2 = 0°$ in fewer iterations than in part (b). **d)** Repeat parts (b) and (c) to show that an adaptive phase-nulling minimum in the SAR at $x = -3$ cm and $x = 3$ cm requires $\phi_2 = 0°$. Hint: Refer to Figure 8.5.

3.14 Show that the average power density expressed by (3.58) reduces to (3.59). Hint: The terms involving the exponential factors integrate to zero. The remaining term does not depend on time.

3.15 Show that average power dissipated according to (3.67) reduces to (3.68).

3.16 Derive (3.109) to (3.114) from (3.102) to (3.108).

References

[1] Field, S.B. and J.W. Hand, (eds.), *An Introduction to the Practical Aspects of Clinical Hyperthermia,* London: Taylor & Francis, 1990.

[2] Gauthrie, M., (ed.), *Methods of External Hyperthermic Heating,* New York, NY: Springer-Verlag, 1990.

[3] Lehmann, J.F., (ed.), *Therapeutic Heat and Cold,* 3rd ed. Baltimore, MD: Williams & Wilkins, 1982.

[4] Duck, F.A., *Physical Properties of Tissue: A Comprehensive Reference Book,* San Diego, CA: Academic Press, 1990.

[5] Woodard, H.Q., and D.R. White, "The Composition of Body Tissues," *The British Journal of Radiology,* Vol. 59, December 1986, pp. 1209-1219.

[6] von Hippel, A.R., *Dielectrics and Waves,* New York: John Wiley, 1954.

[7] Campbell, A.M., and D.V. Land, "Dielectric Properties of Female Human Breast Tissue Measured in Vitro at 3.2 GHz," *Phys Med Biol,* Vol. 37, No. 1, 1992, pp. 193-210.

[8] Foster, K.R., and J.L. Schepps, "Dielectric Properties of Tumor and Normal Tissues at Radio Through Microwave Frequencies," *J Microwave Power,* Vol. 16, No. 2, 1981, pp. 107-119.

[9] England, T.S., and N.A. Sharples, "Dielectric Properties of the Human Body in the Microwave Region of the Spectrum," *Nature,* Vol. 163, March 26, 1949, pp. 477-488.

[10] England, T.S., "Dielectric Properties of the Human Body for Wave-Lengths in the 1-10 cm range," *Nature,* Vol. 166, September 16, 1950, pp. 480-481.

[11] Smith, S.R., and K.R. Foster, "Dielectric Properites of Low-Water-Content Tissues," *Phys Med Biol,* Vol. 30, No. 9, 1985, pp. 965-973.

[12] Chaudhary, S.S., R.K. Mishra, A. Swarup, and J.M. Thomas, "Dielectric Properties of Normal and Malignant Human Breast Tissue at Radiowave and Microwave Frequencies," *Indian J Biochem Biophys,* Vol. 21, 1984, pp. 76-79.

[13] Joines, W.T., Y. Zhang, C. Li, and R.L. Jirtle, "The Measured Electrical Properties of Normal and Malignant Human Tissues From 50 to 900 MHz," *Med Phys,* Vol. 21, No. 4, 1994, pp. 547-550.

[14] Surowiec, A.J., S.S. Stuchly, J.R. Barr, and A. Swarup, "Dielectric Properties of Breast Carcinoma and the Surrounding Tissues," *IEEE Trans Biomed Eng,* Vol. 35, No. 4, 1988, pp. 257-263.

[15] Burdette, E.C. "Electromagnetic and Acoustical Properties of Tissue," In: *Physical Aspects of Hyperthermia,* Nussbaum, G.H., (ed.), AAPM Medical Physics Monographs, No. 8, 1982, pp. 105-130.

[16] Gabriel, S., R.W. Lau, and C. Gabriel, "The Dielectric Properties of Biological Tissues: Part III. Parametric Models for the Dielectric Spectrum of Tissues," *Phys Med Biol,* Vol. 41, 1996, pp. 2271-2293, online http://niremf.ifac.cnr.it/tissprop.

[17] Sha, L., E.R. Ward, B. Story, "A Review of Dielectric Properties of Normal and Malignant Breast Tissue," *Proc IEEE SoutheastCon,* 2002, pp. 457-462.

[18] Lazebnik, M., et al., "A Large-Scale Study of the Ultrawideband Microwave Dielectric Properties of Normal, Benign, and Malignant Breast Tissues Obtained from Cancer Surgeries, *Phys Med Biol,* Vol. 52, 2007, pp. 6093-6115.

[19] Stogryn, A., "Equations for Calculating the Dielectric Constant of Saline Water," *IEEE Trans Microwave Theory and Techniques,* Vol. 19, No. 8, 1971, pp. 733-736.

[20] Malmberg, C.G., and A.A. Maryott, "Dielectric Constant of Water from 0 to 100°C," *J Res National Bureau of Standards,* Vol. 56, No. 1, 1956, pp. 1-8.

[21] Fenn, A.J., *Adaptive Antennas and Phased Arrays for Radar and Communications,* Norwood, MA: Artech House, 2008.

[22] Fenn, A.J., *Breast Cancer Treatment by Focused Microwave Thermotherapy,* Sudbury, MA: Jones and Bartlett, 2007.

[23] Maxwell, J.C., *A Treatise on Electricity and Magnetism,* Oxford, UK: Clarendon Press, Vol. 1 and 2, 1873.

[24] Paris, D.T., and F.K. Hurd, *Basic Electromagnetic Theory,* New York: McGraw-Hill, 1969.

[25] Stratton, J.A., *Electromagnetic Theory,* New York: McGraw-Hill, 1941.

[26] Kraus, J.D., *Electromagnetics,* New York: McGraw-Hill, 1953.

[27] Harrington, R.F., *Time-Harmonic Electromagnetic Fields,* New York: McGraw-Hill, 1961.

[28] Hayt, W.H., *Engineering Electromagnetics,* New York: McGraw-Hill, 1967.

[29] Durney, C.H., and C.C. Johnson, *Introduction to Modern Electromagnetics,* New York: McGraw-Hill, 1969.

[30] Cheng, D.K., *Field and Wave Electromagnetics,* Reading, MA: Addison-Wesley, 1983.

[31] Ramo, S., J.R. Whinnery, and T. van Duzer, *Fields and Waves in Communication Electronics,* New York: John Wiley, 1984.

[32] Sadiku, M.N.O., *Elements of Electromagnetics,* 2nd ed., New York: Oxford University Press, 1995.

[33] Smith, G.S., *An Introduction to Classical Electromagnetic Radiation,* Cambridge, UK: Cambridge University Press, 1997.

[34] Kong, J.A., *Electromagnetic Wave Theory,* Cambridge, MA: EMW Publishing, 2000.

[35] Zhu, Y., and A. Cangellaris, *Multigrid Finite Element Methods for Electromagnetic Field Modeling,* New Jersey: John Wiley, 2006.

[36] Stutzman, W.L., *Polarization in Electromagnetic Systems,* Norwood, MA: Artech House, 1993.

[37] Collin, R.E., *Foundations for Microwave Engineering,* New York: McGraw-Hill, 1966, pp. 64-143.

[38] Kraus, J.D., *Antennas*, (2nd ed.), New York: McGraw-Hill, 1988, pp. 40-42.

[39] Schelkunoff, S.A., and H.T. Friis, *Antennas: Theory and Practice*, New York: Wiley, 1952.

[40] Poynting, J.H., "On the Transfer of Energy in the Electromagnetic Field," *Philosophical Transactions of the Royal Society of London*, Vol. 175, 1884, pp. 343-361.

[41] Lagendijk, J.J.W., and P. Nilsson, "Hyperthermia Dough: A Fat and Bone Equivalent Phantom to Test Microwave/Radiofrequency Hyperthermia Heating Systems," *Phys Med Biol*, Vol. 30, No. 7, 1985, pp. 709-712.

[42] Chou, C-K. G-W. Chen, A.W. Guy, and K.H. Luk, "Formulas for Preparing Phantom Muscle Tissue at Various Radiofrequencies," *Bioelectromagnetics*, Vol. 5, 1994, pp. 435-441.

[43] Lazebnik, M., E.L. Madsen, G.R. Frank, and S.C. Hagness, "Tissue-Mimicking Phantom Materials for Narrowband and Ultrawideband Microwave Applications," *Physics in Medicine and Biology*, Vol. 50, 2005, pp. 4245-4258.

[44] Stauffer, P.R., F. Rossetto, M. Prakash, D.G. Neuman, and T. Lee, "Phantom and Animal Tissues for Modelling the Electrical Properties of Human Liver," *Int J Hyperthermia*, Vol. 19, No. 1, 2003, pp. 89-101.

[45] Richmond, J.H., "Radiation and Scattering by Thin-Wire Structures in the Complex Frequency Domain," The Ohio State University, ElectroScience Laboratory, Technical Report 2902-10, July 1973.

[46] Richmond, J.H., "Computer Program for Thin-Wire Structures in a Homogeneous Conducting Medium," The Ohio State University, ElectroScience Laboratory, Technical Report 2902-12, August 1973.

[47] Richmond, J.H., "Radiation and Scattering by Thin-Wire Structures in a Homogeneous Conducting Medium (Computer Program Description)," *IEEE Trans. Antennas Propagat.*, Vol. AP-22, No. 2, 1974, p. 365.

[48] Richmond, J.H., and N.H. Geary, "Mutual Impedance of Nonplanar-Skew Sinusoidal Dipoles," *IEEE Trans. Antennas Propagat.*, Vol. 25, No. 3, May 1975, pp. 412-414.

[49] Harrington, R.F., *Field Computation by Moment Methods*, New York: The Macmillan Company, 1968.

[50] Stutzman, W.L, and G.A. Thiele, *Antenna Theory and Design*, 2nd ed., New York: Wiley, 1998.

[51] Yee, K.S., "Numerical Solution of Initial Boundary Value Problems Involving Maxwell's Equations in Isotropic Media," *IEEE Trans. Antennas Propagat.*, Vol. 14, No. 3, 1966, pp. 302-307.

[52] Taflove, A., and M.E. Brodwin, "Numerical Solution of Steady-State Electromagnetic Scattering Problems Using the Time-Dependent Maxwell's Equations," *IEEE Trans. Microwave Theory Techniques*, Vol. 23, No. 8, 1975, pp. 623-630.

[53] Taflove, A., and K.R. Umashankar, "Review of FD-TD Numerical Modeling of Electromagnetic Wave Scattering and Radar Cross Section," *Proc. IEEE*, Vol. 77, No.

5, 1989, pp. 682-699.

[54] Taflove, A., (ed.), *Advances in Computational Electrodynamics: The Finite-Difference Time-Domain Method,* Norwood, MA: Artech House, Inc., 1998.

[55] Kunz, K.S., and R.J. Luebbers, *The Finite Difference Time Domain Method,* Boca Raton, Florida: CRC Press, 1993.

[56] Sullivan, D., "Mathematical Methods for Treatment Planning in Deep Regional Hyperthermia," *IEEE Trans on Microwave Theory and Techniques,* Vol. 39, No. 5, 1991, pp. 864-872.

[57] Reuter, C.E., E.T. Thiele, A. Taflove, M.J. Piket-May, and A.J. Fenn, "Linear Superposition of Phased Array Antenna Near-Field Patterns Using the FD-TD Method," *10th Annual Review of Progress in Applied Computational Electromagnetics,* Monterey, California, March 21-26, 1994, pp. 459-466.

4

Thermal Modeling Theory for Tissue Heating

4.1 INTRODUCTION

In developing techniques for electromagnetic hyperthermia treatment of cancer, it is necessary to model the body tissues and antenna applicators and predict the thermal distribution induced in tissue. The previous chapter described the theory for computing the electromagnetic fields generated by an array of antenna applicators surrounding body tissues. In this chapter a theory for computing the time-dependent induced temperature distribution in tissue is reviewed [1-12]. Computation of the tissue temperature distribution requires detailed knowledge of the tissue parameters including specific heat capacity, thermal conductivity, density, metabolic heating rate, and blood flow rate, which are discussed in Section 4.2. The thermal analysis approach, based on the bioheat equation described in Section 4.3, allows for three-dimensional thermal modeling in heterogeneous tissues using the finite-difference technique.

4.2 THERMAL PROPERTIES OF TISSUE

Specific heat capacity of tissue is a measure of the heat energy required to elevate the temperature of a unit quantity of the tissue by a certain temperature increment. Specific heat is typically given in the literature in SI units of J/(g-K) [2, pp. 28-30], which is equivalent to J/(g-°C) or W-s/(g-°C), where J means joules, g is grams, K is Kelvin, and C is Celsius. Specific absorption rate (SAR) in tissue is typically expressed in units of watts per kilogram (W/kg) (refer to (3.69)). Ignoring metabolic heat generation effects, thermal

Table 4.1

Typical Thermal Conductivity, Specific Heat, and Density for Various Tissues

Tissue Type	Thermal Conductivity (W/m - °C)	Specific Heat (kJ/kg - °C)	Density g/cm³
Abdomen (core)	0.544	3.70	1.05
Blood (whole)	0.549	3.64	1.05
Bone (average)	1.16	1.59	1.5
Brain (excised)	0.528	3.68	1.05
Fat (pure)	0.19	2.30	0.85
Heart (excised)	0.586	3.72	1.06
Kidney (excised)	0.544	3.89	1.05
Liver	0.565	3.60	1.05
Muscle (living)	0.642	3.75	1.05
Skin (upper 2 mm)	0.376	3.77	1.00

Source: Lehmann [3, p. 84]

conduction, and blood flow, the rise in temperature in a given time interval is given by the SAR divided by the specific heat. For convenience, specific heat is used here with units of kW-s/(kg-°C), or equivalently 10^3W-s/(kg-°C), or W-s/(10^{-3}kg-°C). As an example, the amount of heat energy to raise one gram of liquid water by one kelvin (same as one degree Celsius) is 4.186 joules per gram or 4.186 watt-seconds per gram.

Typical values for the thermal conductivity, specific heat, and density for various tissues are listed in Table 4.1 [3, p. 84]. The thermal conductivity of the tissues listed in Table 4.1 varies from about 0.19 to 1.16 (W/m-°C). The specific heat of the tissues listed varies from 1.59 to 3.89 (kJ/kg-°C), and the density varies from 0.85 to 1.5 g/cm³. Typical blood flow rates for various tissues [3, p. 89] are given in Table 4.2 and the flow rates are observed to be widely varied depending on the tissue type. For example, blood flow rate is highest in kidney and heart and lowest in bone and fat. Metabolic heat generation rates [3, p. 86] for different tissues are given in Table 4.3. Similar to blood flow rate, metabolic heat generation rate is highest in kidney and heart and lowest in bone and fat.

4.3 BIOHEAT EQUATION

To compute the thermal distribution over a volume, assume a three-dimensional rectangular coordinate system described by the variables (x, y, z). Assume that the tissue can be represented by a lattice with nodes separated by the distances $(\Delta x, \Delta y, \Delta z)$. A simplified diagram showing a uniform node lattice for the x, y plane is shown in Figure 4.1.

Table 4.2
Typical Blood Perfusion Rates for Various Tissues

Tissue Type	Specific Flow Rate (ml/min - 100g)
Bone	0.33 to 0.67
Brain	54.0
Fat (abdomen)	1.84 to 3.04
Heart	84.0
Kidney	420.0
Liver	57.7
Muscle (forearm, resting)	3.2
Skin (abdomen)	8.65
Skin (head)	42.9
Skin (thorax)	6.45

Source: Lehmann [3, p. 89].

Table 4.3
Typical Metabolic Heat Generation Rates for Various Tissues

Tissue Type	Metabolic Heat Generation Rates (W/kg)
Bone	0.051 to 0.0103
Brain	11.0
Fat	0.00405
Heart	33.0
Kidney	20.0
Liver	6.7
Muscle (resting)	0.67
Skin	1.0

Source: Lehmann [3, p. 86].

In hyperthermia treatments, the temperature T induced in tissue at a point (x, y, z) and time t depends on a number of factors that are quantified in terms of the bioheat equation for the time rate of change of temperature with units of °C/s per unit mass as described by Lehmann [3, pp. 77, 225]:

$$\frac{dT(x, y, z, t)}{dt} = \frac{10^{-3}}{c}[W_{SAR} + W_{metab} + W_{cond} - W_{blood}] \qquad (4.1)$$

where, W_{SAR} is the specific absorption rate (SAR), W_{metab} is the metabolic heating rate, W_{cond} is the power dissipated by thermal conduction, and W_{blood} is the power dissipated by blood flow, all expressed in W/kg, c is the specific heat of the tissue in SI units as kJ/(kg-°C) or kW-s/(kg-°C). In some databases,

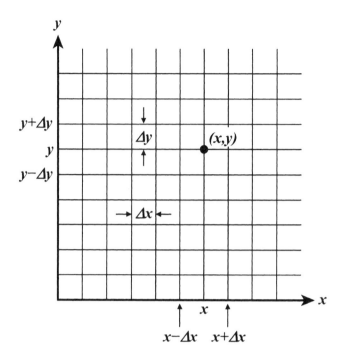

Figure 4.1 Two-dimensional grid for thermal modeling using the finite-difference technique.

the specific heat is given in cgs units as calories/(g-°C), and in 1 joule (or 1 W-s) there are 0.2389 calories, in which case it would be necessary to multiply the right-hand side of (4.1) by the factor 0.2389 [3, p. 77]. An alternate form of the bioheat equation uses specific heat with units of J/(kg-°C) such that (4.1) is expressed in a simpler form as

$$\frac{dT(x,y,z,t)}{dt} = \frac{1}{c}[W_{SAR} + W_{metab} + W_{cond} - W_{blood}] \qquad (4.2)$$

The time derivative on the left side of (4.2) can be computed using finite differences as

$$\frac{dT(x,y,z,t)}{dt} = \frac{T(x,y,z,t+dt) - T(x,y,z,t)}{dt} \qquad (4.3)$$

Thus, (4.2) can be rewritten as

$$T(t+dt) = T(t) + \frac{1}{c}[W_{SAR} + W_{metab} + W_{cond} - W_{blood}]\,dt \qquad (4.4)$$

As described in Chapter 3, the specific absorption rate (SAR) [3, p. 217] is the average power dissipated or absorbed per unit mass (W/kg) of the medium (tissue), or from (3.69) it follows that

$$W_{SAR} = \text{SAR}(x, y, z) = \frac{\sigma}{2\rho} |E(x, y, z)|^2 \tag{4.5}$$

where σ is the electrical conductivity of the medium in S/m, ρ is the density of the medium in kg/m^3, and E is the local electric field with units of volts per meter. The magnitude of the electric field is given in terms of the three rectangular components, E_x, E_y, E_z as

$$|E(x, y, z)|^2 = |E_x|^2 + |E_y|^2 + |E_z|^2 \tag{4.6}$$

The metabolic heating rate is expressed in terms of the baseline body temperature ($37°C$) as [3, p. 225]

$$W_{metab} = W_o(1.1)^{(T(x,y,z,t)-37°C)} \tag{4.7}$$

where, W_o is the initial metabolic heating rate. From (4.7) it is observed that the metabolic heating rate increases as the tissue temperature increases above $37°C$.

The thermal conduction rate can be computed from [3, p. 225]

$$W_{cond} = \frac{k_c}{\rho} \nabla^2 T \tag{4.8}$$

where k_c is the thermal conductivity of the tissue with units of W/(m-K), which is the same as W/(m-$°$C), and ∇^2 is the Laplacian operator.

The Laplacian operator in (4.8) is a second-order differential operator and is defined as the divergence of the gradient in three-dimensional space as

$$\nabla^2 = \nabla \cdot \nabla = \frac{\partial^2}{\partial x^2} + \frac{\partial^2}{\partial y^2} + \frac{\partial^2}{\partial z^2} \tag{4.9}$$

Thus, using (4.9) in (4.8) the thermal conduction rate can be expressed as

$$W_{cond} = \frac{k_c}{\rho} \left[\frac{\partial^2 T}{\partial x^2} + \frac{\partial^2 T}{\partial y^2} + \frac{\partial^2 T}{\partial z^2} \right] \tag{4.10}$$

Referring to Figure 4.2, consider an example of the use of the above equations in calculating the steady-state thermal distribution in tissue, similar to an example described by Lehmann [3, pp. 97-98]. In this simplified

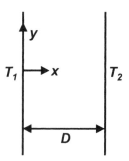

Figure 4.2 Geometry for thermal heating example with tissue sample having thickness D with constant surface temperatures T_1 and T_2.

example, assume that the applied electric field E_o and SAR are uniform throughout the tissue volume, that is,

$$W_{SAR} = SAR = \frac{\sigma}{2\rho}|E_o|^2 \qquad (4.11)$$

It is further assumed that the temperature boundary condition allows only a variation in temperature in the x direction. Ignoring blood perfusion and metabolic heating, in the steady state the one-dimensional thermal equation is expressed as

$$\frac{\partial^2 T}{\partial x^2} = -\frac{\sigma}{2k_c}|E_o|^2 \qquad (4.12)$$

subject to the constant temperature boundary conditions T_1 at $x = 0$ and T_2 at $x = D$. By antidifferentiation of (4.12) it follows that

$$\frac{\partial T}{\partial x} = -\frac{\sigma}{2k_c}|E_o|^2 x + C_1 \qquad (4.13)$$

where C_1 is a constant, and then by antidifferentiation of (4.13) it follows that

$$T(x) = -\frac{\sigma}{4k_c}|E_o|^2 x^2 + C_1 x + C_2 \qquad (4.14)$$

Assume that the surface temperatures are held constant and are equal, that is, $T_1 = T_2 = T_s$. The constants C_1 and C_2 are determined by the boundary conditions; that is,

$$T(x = 0) = T_s = C_2 \qquad (4.15)$$

and then

$$T(x = D) = T_s = -\frac{\sigma}{4k_c}|E_o|^2 D^2 + C_1 D + T_s \qquad (4.16)$$

so

$$C_1 = \frac{\sigma}{4k_c}|E_o|^2 D \tag{4.17}$$

Therefore, for this example, the one-dimensional thermal distribution is expressed as

$$T(x) = \frac{\sigma}{4k_c}|E_o|^2(D - x)x + T_s \tag{4.18}$$

To determine the peak value, take the derivative of (4.18) and equate to zero, that is,

$$T'(x) = \frac{\sigma}{4k_c}|E_o|^2(D - 2x) = 0 \tag{4.19}$$

and thus the peak occurs when $x = D/2$, or the midpoint of the tissue volume as expected by symmetry in this example.

For the general case, the partial derivatives in (4.10) can be evaluated numerically by using a Taylor's expansion [12, p. 483], [13, p. 84], with one of the dimensions variable and the other two dimensions fixed. For example, as dx goes to zero

$$T(x + dx) = T(x) + dx\frac{dT(x)}{dx} + \frac{dx^2}{2}\frac{d^2T(x)}{dx^2} + \cdots \tag{4.20}$$

$$T(x - dx) = T(x) - dx\frac{dT(x)}{dx} + \frac{dx^2}{2}\frac{d^2T(x)}{dx^2} - \cdots \tag{4.21}$$

Adding the above two equations and ignoring terms higher than second order yields

$$T(x + dx) + T(x - dx) = 2T(x) + dx^2\frac{d^2T(x)}{dx^2} \tag{4.22}$$

Solving for the second derivative yields

$$\frac{d^2T}{dx^2} = \frac{T(x + dx) - 2T(x) + T(x - dx)}{dx^2} \tag{4.23}$$

and similarly it follows that

$$\frac{d^2T}{dy^2} = \frac{T(y + dy) - 2T(y) + T(y - dy)}{dy^2} \tag{4.24}$$

$$\frac{d^2T}{dz^2} = \frac{T(z + dz) - 2T(z) + T(z - dz)}{dz^2} \tag{4.25}$$

The thermal conduction rate can be calculated by substituting (4.23) to (4.25) into (4.10) The finite-difference technique can be used to solve the above differential equations by replacing (dx, dy, dz, dt) with $(\Delta x, \Delta y, \Delta z, \Delta t)$

appropriately. For example, in the one-dimensional case involving the variable x, the thermal conduction rate can be computed numerically as

$$W_{cond} = \frac{k_c}{\rho} \frac{T(x + \Delta x) - 2T(x) + T(x - \Delta x)}{(\Delta x)^2} \qquad (4.26)$$

To include the effects of blood perfusion, the power dissipated by blood flow is expressed as [3, p. 225]

$$W_{blood} = k_2 m c_b \rho_b (T - T_a) \qquad (4.27)$$

where the constant $k_2 = 0.698$, m is the blood flow rate with units of ml/min-100 g, c_b is the specific heat of blood, ρ_b is the density of blood, and blood enters the tissue at arterial temperature T_a.

In the absence of applied power, the equilibrium values of terms in (4.2) are on the order of 1 W/kg for typical resting muscle. To elevate tissue temperature at a reasonable rate, the SAR must be on the order of 50 W/kg or greater [3, p. 225].

Phantom testing with simulated tissues is often conducted by microwave heating over a short period of time, usually with microwave power bursts of 15 to 30 seconds for each heating period. For short-burst phantom testing, the metabolic rate, thermal conduction, and blood perfusion can be ignored in (4.4), and it follows that,

$$T(t + dt) = T(t) + \frac{1}{c}\left[SAR(x, y, z)\right] dt = \frac{1}{c}\left[\frac{\sigma}{2\rho}|E(x, y, z)|^2\right] dt \quad (4.28)$$

or using $\Delta T = T(t + dt) - T(t)$ and $\Delta t = dt$ in (4.28) yields

$$SAR = c\frac{\Delta T}{\Delta t} = \frac{1}{2}\frac{\sigma}{\rho}|E(x, y, z)|^2 \qquad (4.29)$$

Thus, it follows from (4.29) that for a short-burst time interval Δt of heating, that the rise in temperature of a phantom is given by

$$\Delta T = \frac{1}{c}SAR\Delta t \qquad (4.30)$$

4.4 SUMMARY

This chapter has reviewed a mathematical formulation for the analysis of the spatial and time-dependent temperature induced in human tissues by electromagnetic fields. The formulation makes use of the finite-difference method to solve the bioheat equation. Thermal parameters of typical tissues have been given based on data in the literature. The next chapter applies the formulation described here for predicting the thermal distribution for an adaptive phased array thermotherapy system irradiating tissue.

References

[1] Pennes, H.H., Analysis of Tissue and Arterial Blood Temperatures in the Human Resting Forearm, *J Applied Physiology*, Vol. 1, No. 2, August 1948, pp. 93-122.

[2] Duck, F.A., *Physical Properties of Tissue, A Comprehensive Reference Book*, San Diego: Academic Press, 1990.

[3] Lehmann, J.F., (ed.), *Therapeutic Heat and Cold*, 3rd ed., Baltimore: Williams & Wilkins, 1982.

[4] Mooibroek, J., J. Crezee, and J.J.W. Lagendijk, "Basics of Thermal Models," In: Chapter 19 of *Thermoradiotherapy and Thermochemotherapy, Vol. 1, Biology, Physiology, and Physics,*, Seegenschmiedt, M.H., P. Fessenden, and C.C. Vernon, (eds.), Berlin: Springer-Verlag, 1995, pp. 425-437.

[5] Guy, A.W., "History of Biological Effects and Medical Applications of Microwave Energy," *IEEE Trans on Microwave Theory and Techniques,* Vol. MTT-32, No. 9, 1984, pp. 1182-1200.

[6] Bowman, H.F., "Heat Transfer and Thermal Dosimetry," *J Microwave Power,* Vol. 16, No. 2, June 1981, pp. 121-133.

[7] Moros, E.G, A.W. Dutton, R.B. Roemer, M. Burton, and K. Hynynen, "Experimental Evaluation of Two Simple Thermal Models Using Hyperthermia in Muscle in Vivo," *Int J Hyperthermia*, Vol. 9, No. 4, 1993, pp. 581-598.

[8] Chen, C., and R.B. Roemer, "A Thermo-Pharmacokinetic Model of Tissue Temperature Oscillations During Localized Heating," *Int J Hyperthermia,* Vol. 21, No. 2, 2005, pp. 107-124.

[9] Clegg, C.T., S.K. Das, Y. Zhang, J. MacFall, E. Fullar, T.V. Samulski, "Verification of a Hyperthermia Model Using MR Thermometry," *Int J Hyperthermia,* Vol. 11, No. 3, 1995, pp. 409-424.

[10] Zhang, Y., W.T. Joines, J.R. Oleson, "The Calculated and Measured Temperature Distribution of a Phased Interstitial Antenna Array," *IEEE Trans on Microwave Theory and Techniques*, Vol. 38, No. 1, 1990, pp. 69-77.

[11] Lagendijk, J.J.W., "Thermal Models: Principles and Implementation," In: *An Introduction to the Practical Aspects of Clinical Hyperthermia,* Field, S.B. and J.W. Hand, (eds.), London: Taylor & Francis, 1990, pp. 478-512.

[12] Field, S.B. and J.W. Hand, (eds.), *An Introduction to the Practical Aspects of Clinical Hyperthermia,* London: Taylor & Francis, 1990.

[13] Ralston, A., *A First Course in Numerical Analysis,* New York: McGraw-Hill, 1965, p. 84.

5

Adaptive Array Simulations for the Torso

5.1 INTRODUCTION

In this chapter, the adaptive phased array approach [1-25] discussed in Chapters 1 and 2 is applied to the problem of generating a therapeutic thermal dose distribution in electromagnetic hyperthermia treatment of cancer. With this minimally invasive adaptive phased array system approach, it may be possible to maximize the applied electric field at a tumor position in a target body and simultaneously minimize or reduce the electric field at target positions where undesired high temperature regions (hot spots) occur. An adaptive transmit phased array is analyzed here for a simple homogeneous target case.

As described in Chapter 2, in a preclinical or clinical setting, either a gradient-search algorithm or sample matrix inversion algorithm could be used to rapidly form the desired adaptive null (or nulls) prior to any significant tissue heating. In this chapter, the sample matrix inversion algorithm is used to form adaptive nulls. Auxiliary short-dipole field probes are used in sampling the local elecric-field amplitude and in effecting the desired electric-field nulls. A radiofrequency adaptive null formed at the surface of the target has a finite width and extends into the target. This finite null width can allow for noninvasively positioned auxiliary probes in an adaptive hyperthermia system. The allowed spacing or resolution between a deep null and focus is fundamentally equal to the hyperthermia antenna half-power beamwidth in the tissue. A closer spacing between the desired null position and focus can be achieved by limiting the null depth.

As will be demonstrated in this chapter, analysis of an adaptive transmit phased array antenna system shows the potential merit of combining adaptive nulling with near-field focusing used in hyperthermia. The simplified three-dimensional analysis used here is based on a well-known moment-method theory for conducting thin-wire antennas in a homogeneous conducting medium as described in Chapter 3. The theory is used to compute the moment-method received voltages at a set of short-dipole auxiliary probes, due to a transmitting dipole array. Computer simulation results are presented for a fully adaptive eight-element annular phased array operating at a radiofrequency of 120 MHz. Multiple simultaneous nulls are used in adaptively generating a desired radiation pattern in simulated body tissue. The power received at a movable short-dipole E-field sensor is then used as the RF power source in a second simulation that performs a transient thermal analysis, according to the theory described in Chapter 4, for an elliptical target body surrounded by a constant-temperature water bolus. The computer simulations show that noninvasive adaptive nulling can prevent undesired high-temperature regions from occurring while simultaneously heating a deep-seated tumor site.

5.2 SIMULATION MODEL

To demonstrate the potential effectiveness of focused near-field adaptive nulling in reducing undesired hot spots, computer simulations are presented in this chapter. The E-field simulations are based on the signal received by a short-dipole probe due to a transmitting phased array embedded in an infinite homogeneous lossy dielectric (muscle tissue) with a defined elliptical target region. Because the RF dielectric constant and wavelength in the elliptical target and surrounding water bolus are similar, the E-field simulations provide a reasonable approximation to the field distribution inside the elliptical target, as was observed in Chapter 1 in the comparison of moment-method and finite-difference time-domain simulations shown in Figure 1.12. To describe this comparison further, Sullivan [26] has analyzed the performance of an eight-element dipole ring array for hyperthermia, and has shown finite-difference time-domain simulations and measurements of SAR patterns at 70 to 110 MHz, using a homogeneous 35 × 25-cm elliptical target surrounded with a layer of fat and a water bolus. A comparison of Sullivan's 110-MHz SAR simulation (Figure 5(a) of his paper) and method of moments simulated results by this author, previously shown in Figure 1.12, indicates good agreement over the elliptical target region. Having shown the validity of analyzing a hyperthermia phased array by this method; that is, the simple moment-method model for approximating the electric-field distribution in a homogeneous

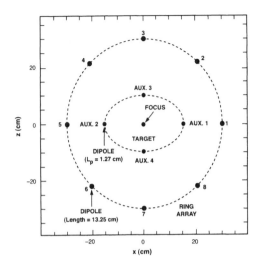

Figure 5.1 Geometry for eight-element ring array of transmit dipoles and four E-field auxiliary nulling dipole sensors surrounding a target zone.

target, the next step is to compute the fields and thermal distribution generated by an adaptive phased array. The calculated field distribution is used as the power source in a thermal simulation for an elliptical target (muscle tissue) surrounded by a constant-temperature water bolus.

The computer simulation model is related, in part, to the hyperthermia annular dipole phased array antenna system shown in Chapter 6 (Figures 6.2 and 6.3). This hyperthermia system uses a 60-cm array diameter with eight uniformly spaced dipole elements (fed as four pairs) that operate over the frequency band 60 to 120 MHz [27, 28]. The adaptive phased array system analyzed here has eight radiating dipole antenna elements and four auxiliary receive antennas, as depicted in Figure 5.1. The array simulated here is assumed to be fully adaptive, and with eight adaptive array elements, up to seven independent nulls can be formed while simultaneously focusing on a tumor. Likewise, a four-channel adaptive phased array hyperthermia system could be used to form up to three independent nulls.

In this chapter, it is assumed that the adaptive radiation pattern null-width characteristics in a homogeneous target will be similar to the characteristics observed in an inhomogeneous target. The null width is directly related to the wavelength and, as will be shown, there is only a 5% change in wavelength between the assumed muscle tissue and water bolus. With this assumption, the transmit array is embedded in homogeneous tissue, which allows direct use of the thin-wire moment-method formulation reviewed in Chapter 3. The electric-field distribution in the homogeneous medium is computed, and then just an elliptical portion of the homogeneous region is used as the homogeneous target. In the thermal analysis, the elliptical target is surrounded with a constant-temperature water bolus. The electric-field amplitude is scaled to produce a 46°C peak temperature, at time $t = 20$ minutes, at the center of the elliptical phantom. The initial temperature of the phantom is assumed to be 25°C (room temperature).

The computer simulations presented in this chapter are for a 120-MHz radiofrequency with four auxiliary nulling sensors; that is, $N_{aux} = 4$ (refer to Figures 1.10 and 5.1). The four nulling sensors are placed anterior, posterior, lateral left, and lateral right. The reason for choosing these nulling sensor positions is as follows. In the example that follows, it is assumed that the desired focal position is centrally located at maximum depth in the target. If the nulling sensor positions were to favor one side of the target, the focal beam could shift in the opposite direction when the adaptive nulls are formed. By using a symmetrical arrangement of nulling sensors, it is shown that it is possible to maintain a centrally deep focus.

The parameters used in the electrical and thermal analyses are summarized in Table 5.1. Notice that the relative dielectric constants of phantom muscle tissue and distilled water are very similar; however, the electrical conductivities are vastly different. The relevant thermal characteristics—density, specific heat, and thermal conductivity [29]—are very similar for phantom muscle tissue and distilled water.

From Table 5.1, substituting the values $\sigma = 0.5$ S/m and $\epsilon_r = 73.5$ into (3.32), with $f = 120$ MHz yields a propagation constant of $\gamma_m = 10.0 + j23.8$ (with units of nepers per meter and radians per meter for the real and imaginary components, respectively) for the phantom muscle tissue. With the phase constant $\beta_m = 23.8$ radians/m, the wavelength in the phantom muscle tissue is computed from (3.36) to be $\lambda_m = 26.5$ cm. The attenuation constant

Table 5.1
Parameters Used in Electrical/Thermal Analysis

Parameter	Phantom Muscle Tissue	Distilled Water
Dielectric Constant at 120 MHz	73.5	80.0
Electrical Conductivity at 120 MHz	0.5 S/m	0.0001 S/m
Density	0.97 g/cm^3	1.0 g/cm^3
Specific Heat	3.5 kJ/kg °C	4.2 kJ/kg °C
Thermal Conductivity	0.544 W/m °C	0.60 W/m °C

for the phantom muscle tissue is $\alpha_m = 10.0$ nepers/m. Similarly, for distilled water the propagation constant is $\gamma_w = 0.0021 + j22.5$, so the wavelength is $\lambda_w = 27.9$ cm. The attenuation constant for the distilled water medium is $\alpha_w = 0.0021$ nepers/m. The propagation loss in the phantom muscle tissue is $20 \log_{10} e^{-10.0}$, or -0.87 dB/cm. Note, as discussed in Chapter 3, the propagation loss in decibels per length can be calculated by multiplying the attenuation constant in nepers/length by 8.686. Similarly, the propagation loss in the distilled water is found to be -0.0002 dB/cm. Thus, the total loss due to propagation through 15 cm of distilled water is 0.003 dB, which is negligible. For 15 cm of the phantom muscle tissue the corresponding loss is 13.1 dB. The wave impedance in the phantom muscle tissue is computed from (3.37) as $\eta_m = 33.9 + j14.2$ ohms, with magnitude $|\eta_m| = 36.75$ ohms, and similarly in the distilled water $\eta_w = 42.1 + j0.004$ ohms, with magnitude $|\eta_w| = 42.1$ ohms.

The geometry used in the simulations was shown in Figure 5.1. A 60-cm-diameter uniform ring array of eight dipoles surrounds a target zone defined as a torso-shaped ellipse with major axis 30 cm and minor axis 20 cm. The length of each perfectly conducting center-fed dipole array element is 13.25 cm, which corresponds to one-half wavelength in the simulated tissue at 120 MHz. The array focus is assumed at the origin, and four auxiliary short-dipole sensors with length 1.27 cm (0.05λ) are positioned in (x, y, z) coordinates at (15 cm, 0, 0), $(-15$ cm, 0, 0), (0, 0, 10 cm), and (0, 0, -10 cm); that is, the auxiliary E-field sensors are located every $90°$ in azimuth on the perimeter of the target. The auxiliary element 1, 2, 3, and 4 correspond to the lateral right, lateral left, anterior, and posterior locations, respectively. In rectangular coordinates, each dipole is oriented in the $\hat{y}$ direction and the feed terminals of each dipole are located at $y = 0$. The null depth at the E-field sensors is proportional to the SNR at the auxiliary probe positions. The E-field data that follow are computed for the case of a 1.27-cm short-dipole observation probe.

A commericial software program called TTA, described by Bartoszek [30], was used to implement the thermal analysis described in Chapter 4 to compute the temperature distribution in an elliptical phantom surrounded with a constant-temperature water bolus.

5.3 SIMULATED RESULTS

For the adaptive phased array simulations shown here, the eight-element dipole ring array shown in Figure 5.1 was initially focused at the origin of the defined elliptical target. Next, adaptive radiation patterns were computed with four auxiliary dipole sensors; their positions were shown in Figure 5.1. The

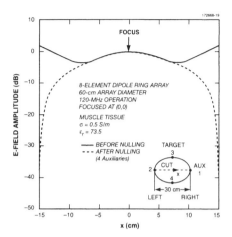

Figure 5.2 Simulated one-dimensional quiescent and adaptive radiation patterns in the $z = 0$ cut (major axis) at 120 MHz for the eight-element ring array in infinite homogeneous conducting medium (phantom muscle tissue: $\epsilon_r = 73.5, \sigma = 0.5$).

receiving gain for auxiliary dipoles 1 and 2 is adjusted to produce a greater-than-35-dB SNR. This SNR level results in greater than 35 dB of nulling in the direction of auxiliary dipoles 1 and 2. In contrast, the gain values for auxiliary dipoles 3 and 4 are adjusted to produce about a 3-dB SNR. Thus, only about 3 dB of nulling will occur at sensor positions 3 and 4 as the adaptive algorithm reduces the interference to the noise level of the receiver.

Calculated radiation patterns before and after adaptive nullling along the major axis (at $z = 0$) of the target are shown in Figure 5.2. Similarly, Figure 5.3 shows the one-dimensional radiation patterns, along the minor axis of the target, before and after nulling. Before nulling (solid curve in Figure 5.2), a focused peak E-field is formed at the origin; however, the larger E-field amplitude sensed at ±15 cm is due to the E-field probe's close proximity to the transmitting elements. The field attenuation that occurs in moving from ±15 cm toward the focus is due to the 1/r attenuation loss and the loss in the uniform homogeneous muscle tissue. From Figure 5.2, the ring-array half-power beamwidth in the target region is observed to be approximately 13 cm, or approximately one-half the wavelength (26.5 cm) in the phantom muscle tissue. The adaptive nulling resolution or closest allowed spacing between a deep adaptive null and the main beam is equal to the half-power beamwidth of the antenna array. Thus, the closest allowed null position is approximately 13 cm from the focus. Since the target width is 30 cm, in theory two nulls can be formed at ($x = \pm15$ cm, $z = 0$) without disturbing the focus. However, if two deep nulls are formed at ($x = 0, z = \pm10$ cm)

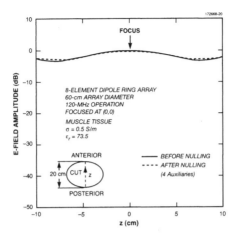

Figure 5.3 Simulated one-dimensional quiescent and adaptive radiation patterns in the $x = 0$ cut (minor axis) at 120 MHz for the eight-element ring array in infinite homogeneous conducting medium (phantom muscle tissue: $\epsilon_r = 73.5, \sigma = 0.5$).

the focus could be compromised. In practice, a water bolus would limit the placement of short-dipole sensors to the surface of the target. Thus, only weak nulls are formed at ($x = 0, z = \pm10$ cm) so that the focus will not be affected by the adaptive nulling process. Two deep adaptive nulls at $x = \pm15$ cm occur as expected due to the strong SNR, and weak nulling occurs at $z = \pm10$ cm, also as expected due the weaker SNR. In the $z = 0$ pattern plot, greater than 35 dB of interference nulling or pattern reduction occurs at $x = \pm15$ cm. The peak level at the focus is normalized to 0 dB for both the quiescent and adaptive patterns. Two weak adaptive nulls are in effect in the $x = 0$ radiation pattern plot shown in Figure 5.3; however, weak nulls are desired in this pattern cut due to temperature considerations, as shown in the next section. The weak nulls in effect in the adaptive patterns reduce variation from the quiescent radiation pattern.

The transmit-array weights before and after nulling and the covariance matrix eigenvalues (degrees of freedom consumed) are shown in Figures 5.4 and 5.5, respectively. The adaptive transmit weights exhibit a 5-dB dynamic range in Figure 5.4(a). There are two large eigenvalues and two weak (nonzero) eigenvalues shown in Figure 5.6. These eigenvalues are directly associated with the two high-SNR auxiliary sensors and the two weak-SNR auxiliary sensors. Note that the 0-dB level in Figure 5.5 is equal to the noise level. The probe-array output power before and after adaptive nulling is 31.4 dB and 0.9 dB, respectively. This difference in power before and after

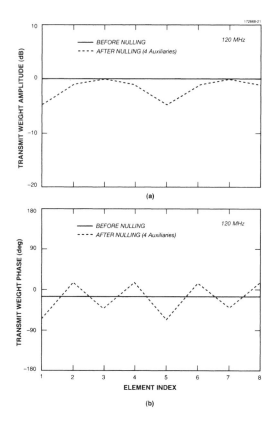

Figure 5.4 Transmit-array weights before and after adaptive nulling. Four auxiliary sensors are used in the adaptive phased array process. (a) Amplitude; (b) phase.

nulling means that the adaptive cancellation is 30.5 dB. Two-dimensional E-field radiation pattern data (at 120 MHz) on a 41×41 grid are used as the power source for the thermal node network. A node spacing $\Delta x = \Delta z = \Delta l = 0.9525$ cm was used. A scale factor was used to convert the normalized E-field distributions to a power level that induces a $46°C$ peak temperature at $t = 20$ minutes for the quiescent and adaptive patterns. The parameter values used in the thermal simulation are listed in Table 5.1. A constant temperature of $10°C$ is enforced at each water-bolus node. Two-dimensional thermal distributions before and after nulling are shown in Figures 5.6 and 5.7, respectively. Similarly, one-dimensional thermal pattern plots are shown in Figures 5.8 (major axis) and 5.9 (minor axis). These simulations show that hot spots away from the focus are at a $42°C$ level before nulling (Figures 5.6 and 5.8). After adaptive nulling, the hot spots are eliminated (Figures 5.7 and

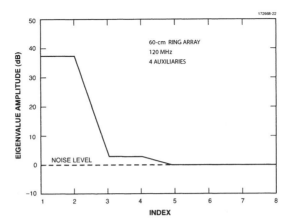

Figure 5.5 Channel covariance matrix eigenvalues (degrees of freedom) used in the adaptive phased array process with four auxiliary sensors.

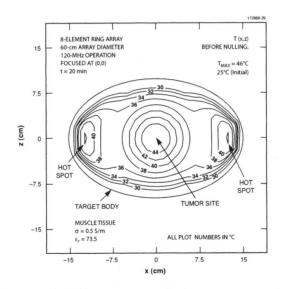

Figure 5.6 Simulated two-dimensional thermal pattern (before adaptive nulling) in elliptical phantom muscle-tissue target surrounded with a water bolus. The quiescent incident RF power distribution is at 120 MHz. Hot spots on the left and right sides of the target are present. (From [13] with permission from Informa Healthcare, www.informaworld.com.)

5.8). Figure 5.9 shows that along the minor axis, there are no hot spots before or after adaptive nulling.

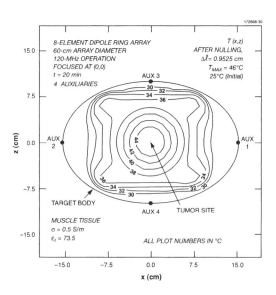

Figure 5.7 Simulated two-dimensional thermal pattern (with adaptive nulling in effect) in elliptical phantom muscle-tissue target surrounded with a water bolus. Hot spots on the left and right sides of the target are eliminated, as previously observed before nulling in Figure 5.6. (From [13] with permission from Informa Healthcare, www.informaworld.com.)

5.4 SUMMARY

In this chapter, an adaptive phased array system for deep tumor treatments has been analyzed. The electric-field distribution for an annular phased array irradiating tissue has a large amplitude in moving away from the central focal position toward the defined body perimeter. This large amplitude gives rise to lateral hot spots. By adaptively nulling the electric field on the target perimeter, the electric field was reduced over an interior region adjacent to the target perimeter. The noninvasively formed null has a finite width and extends into the target region. The centrally focused beam was maintained in the adaptive nulling process by use of multiple auxiliary probes on the target perimeter. Thermal analysis showed lateral hot spots away from the central focus prior to adaptive nulling. After adaptive nulling, the lateral hot spots were eliminated. The next chapter provides experimental testing in phantoms that demonstrate focused near-field adaptive nulling for deep hyperthermia.

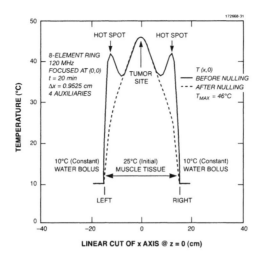

Figure 5.8 Simulated one-dimensional thermal patterns along the major axis ($z = 0$) at time $t = 20$ minutes before and after nulling in elliptical phantom muscle-tissue target surrounded with a constant-temperature water bolus. Hot spots on the left and right sides of the target are eliminated by the adaptive nulling process.

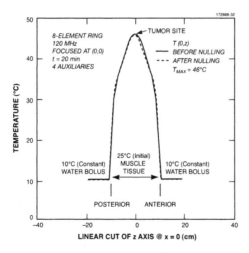

Figure 5.9 Simulated one-dimensional thermal patterns along the minor axis ($x = 0$) at time $t = 20$ minutes before and after nulling in elliptical phantom muscle-tissue target surrounded with a constant-temperature water bolus. No undesired hot spots are present.

References

[1] Fenn, A.J., et al., "The Development of Phased Array Radar Technology," *Lincoln*

Laboratory Journal, Vol. 12, No. 2, 2000, pp. 321-340.

[2] Fenn, A.J., *Adaptive Antennas and Phased Arrays for Radar and Communications*, Norwood, MA: Artech House, 2008.

[3] Fenn, A.J., "Theory and Analysis of Near Field Adaptive Nulling," *1986 IEEE Antennas and Propagation Symposium Digest, Vol. 2*, New York: 1986, pp. 579-582.

[4] Fenn, A.J., "Theory and Analysis of Near Field Adaptive Nulling," *Proc Asilomar Conf Signals, Systems and Computers,* Computer Society Press of the IEEE, Washington, D.C.: Nov 10-12, 1986, pp. 105-109.

[5] Fenn, A.J., "Evaluation of Adaptive Phased Array Far-Field Nulling Performance in the Near-Field Region," *IEEE Trans. Antennas Propagat.*, Vol. 38, No. 2, 1990, pp. 173-185.

[6] Fenn, A.J., H.M. Aumann, F.G. Willwerth, and J.R. Johnson, "Focused Near-Field Adaptive Nulling: Experimental Investigation," *1990 IEEE Antennas Propagation Soc. Int. Symp. Digest,* Vol. 1, May 7-11, 1990, pp. 186-189.

[7] Fenn, A.J., "Analysis of Phase-Focused Near-Field Testing for Multiphase-Center Adaptive Radar Sysems," *IEEE Trans. Antennas Propagat.*, Vol. 40, No. 8, 1992, pp. 878-887.

[8] Fenn, A.J., "Moment Method Analysis of Near Field Adaptive Nulling," *IEE Sixth Int. Conf. on Antennas and Propagation, ICAP 89,* April 4-7, 1989, pp. 295-301.

[9] Fenn, A.J., "Adaptive Nulling Hyperthermia Array," US Patent No. 5,251,645, October 12, 1993.

[10] Fenn, A.J., "Adaptive Focusing and Nulling Hyperthermia Annular and Monopole Phased Array Applicators," US Patent No. 5,441,532, August 15, 1995.

[11] Fenn, A.J., "Non-Invasive Adaptive Nulling for Improved Hyperthermia Thermal Dose Distribution," *IEEE Engineering in Medicine and Biology Society Int Conf,* October 31 - November 3, 1991, Vol. 13, No. 2, 1991, pp. 976-977.

[12] Fenn, A.J., and G.A. King, "Experimental Investigation of an Adaptive Feedback Algorithm for Hot Spot Reduction in Radio-Frequency Phased-Array Hyperthermia," *IEEE Trans Biomed Eng.*, Vol. 43, No. 3, 1994, pp. 273-280.

[13] Fenn, A.J., and G.A. King, "Adaptive Radio Frequency Hyperthermia Phased Array System for Improved Cancer Therapy: Phantom Target Measurements," *Int J Hyperthermia*, Vol. 10, No. 2, 1994, pp. 189-208.

[14] Fenn, A.J., B.A. Bornstein, G.K. Svensson, and H.F. Bowman, "Minimally Invasive Monopole Phased Arrays for Hyperthermia Treatment of Breast Carcinomas: Design and Phantom Tests," *Int. Symp. on Electromagnetic Compatibility*, Sendai, Japan: Vol. 10, No. 2, 1994, pp. 566-569.

[15] Fenn, A.J., "Minimally Invasive Monopole Phased Arrays for Hyperthermia Treatment of Breast Cancer," In: *Proc. 1994 Int. Symp. on Antennas,* Nice, France: November 8-10, 1994, 418-421.

[16] Fenn, A.J., "Minimally Invasive Monopole Phased Array Hyperthermia Applicators and Method for Treating Breast Carcinomas," US Patent No. 5,540,737, July 30, 1996.

[17] Sathiaseelan, V., A.J. Fenn, and A. Taflove, "Recent Advances in External Electromagnetic Hyperthermia," In: Chapter 10 of *Advances in Radiation Treatment*, Mittal, B.B., J.A. Purdy, and K.K. Ang, (eds.), Boston, Massachusetts: Kluwer Academic Publishers, 1998, pp. 213-245.

[18] Fenn, A.J., V. Sathiaseelan, G.A. King, and P.R. Stauffer, "Improved Localization of Energy Deposition in Adaptive Phased Array Hyperthermia Treatment of Cancer," *J Oncol Management*, Vol. 7, No. 2, 1998, pp. 22-29.

[19] Fenn, A.J., "Thermodynamic Adaptive Phased Array System for Activating Thermosensitive Liposomes in Targeted Drug Delivery," US Patent No. 5,810,888, September 22, 1998.

[20] Fenn, A.J., G.L. Wolf, and R.M. Fogle, "An Adaptive Phased Array for Targeted Heating of Deep Tumors in Intact Breast: Animal Study Results," *Int J Hyperthermia*, Vol. 15, No. 1, 1999, pp. 45-61.

[21] Gavrilov, L.R., J.W. Hand, J.W. Hopewell, and A.J. Fenn, "Pre-clinical Evaluation of a Two-Channel Microwave Hyperthermia System with Adaptive Phase Control in a Large Animal," *Int J Hyperthermia*, Vol. 15, No. 6, 1999, pp. 495-507.

[22] Gardner, R.A., H.I. Vargas, J.B. Block, C.L. Vogel, A.J. Fenn, G.V. Kuehl, and M. Doval, "Focused Microwave Phased Array Thermotherapy for Primary Breast Cancer," *Ann Surg Oncol*, Vol. 9, No. 4, 2002, pp. 326-332.

[23] Vargas, H.I., W.C. Dooley, R.A. Gardner, K.D. Gonzalez, S.H. Heywang-Kobrunner, and A.J. Fenn, "Focused Microwave Phased Array Thermotherapy for Ablation of Early-Stage Breast Cancer: Results of Thermal Dose Escalation," *Ann Surg Oncol*, Vol. 11, No. 2, 2004, pp. 139-146.

[24] Fenn, A.J., *Breast Cancer Treatment by Focused Microwave Thermotherapy*, Sudbury, MA: Jones and Bartlett, 2007, pp. 54-56.

[25] Vargas, H.I., W.C. Dooley, A.J. Fenn, M.B. Tomaselli, and J.K. Harness, "Study of Preoperative Focused Microwave Phased Array Thermotherapy in Combination With Neoadjuvant Anthracycline-Based Chemotherapy for Large Breast Carcinomas," *Cancer Therapy*, Vol. 5, 2007, pp. 401-408, published online (www.cancer-therapy.org), November 25, 2007.

[26] Sullivan, D., "Mathematical Methods for Treatment Planning in Deep Regional Hyperthermia," *IEEE Trans. Microwave Theory and Techniques*, Vol. 39, No. 5, 1991, pp. 864-872.

[27] Turner, P.F., T. Schaefermeyer, and T. Saxton, "Future Trends in Heating Technology of Deep-Seated Tumors," *Recent Results in Cancer Research*, Vol. 107, 1988, pp. 249-262.

[28] Turner, P.F., A. Tumeh, and T. Schaefermeyer, "BSD-2000 Approach for Deep Local and Regional Hyperthermia: Physics and Technology," *Strahlentherapie Onkologie*, Vol. 165, No. 10, 1989, pp. 738-741.

[29] Zhang, Y., W.T. Joines, and J.R. Oleson, "The Calculated and Measured Temperature Distribution of a Phased Interstitial Antenna Array," *IEEE Trans on Microwave Theory and Techniques*, Vol. 38, No. 1, 1990, pp. 69-77.

[30] Bartoszek, J.T., B. Huckins, M. Coyle, "A Simplified Shuttle Payload Thermal Analyzer (SSPTA) Program," *AIAA 14th Thermophysics Conference*, Orlando, Florida: June 4-6, 1979, paper 79-1052.

6

Phantom Studies for Deep Tumors in the Torso

6.1 INTRODUCTION

Computer simulations shown in Chapter 5 suggest that deep tumor therapy treatments with an adaptive phased array might reduce hot spots by means of noninvasive adaptive nulling. The simulations in Chapter 5 were limited to a homogeneous tissue target. In this chapter, a computer-controlled adaptive phased array radiofrequency hyperthermia system for improved therapeutic tumor heating is experimentally investigated in homogeneous and heterogeneous phantoms. Referring to the conceptual diagram shown in Figure 6.1, adaptive phased array nulling and focusing techniques [1-28] are used to modify the electric-field distribution in the target body. The example nulls shown in Figure 6.1 each have a width that penetrates the body and can be used to protect healthy tissues away from the tumor target. Figure 5.2 showed an example of the finite null width effect on the electric field, and Figures 5.6 and 5.8 showed the effect on the thermal distribution in a homogeneous target. This chapter shows measured results for adaptive phased arrays irradiating homogeneous and heterogeneous phantom targets.

A hyperthermia phased array antenna system for deep torso heating has been used in demonstrating adaptive phased array gradient-search algorithms [12, 14, 15, 19-21] that were described in Chapter 2. The hyperthermia system investigated in this chapter is a dipole phased array antenna applicator, modified for these experiments, with four independently controlled RF transmitter channels operating at a continuous wave (CW) frequency in the range of 100 to 120 MHz. The hyperthermia phased array is made adaptive by software that invokes a gradient-search feedback algorithm that controls

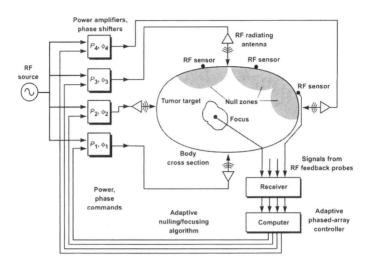

Figure 6.1 A minimally invasive adaptive phased array system of multiple coherent electromagnetic radiating antennas is used in heating a deep-seated tumor, while nulling the electric field at multiple surface positions. At RF frequencies around 100 MHz, the nulls formed on the surface are sufficiently broad such that nearby regions of healthy tissue are protected from RF irradiation [19].

the amplitude and phase of each transmitter channel. The gradient-search algorithm implements the method of steepest descent for adaptive nulling (power minimization) and the method of steepest ascent for adaptive focusing (power maximization). The feedback signals are measured by electric-field short-dipole probe antennas. The measured data indicate that, with an adaptive phased array hyperthermia system, it may be possible to maximize the applied electric field at a tumor position in a complex-scattering heterogeneous target body, and simultaneously minimize or reduce the electric field at target positions where undesired high-temperature regions (hot spots) occur.

The remainder of this chapter is organized as follows. The next section describes the materials and methods used in obtaining the measured results. The phantom targets measured in this chapter are a saline-filled cylindrical phantom, a heterogeneous beef phantom, and a light-emitting diode saline-filled elliptical phantom. Section 6.3 gives measured results for an adaptive four-channel hyperthermia dipole phased array operating at CW frequencies of 100 and 120 MHz. The measured received RF-power distributions before and after adaptive nulling and focusing with short-dipole field probes are presented. Section 6.4 presents the conclusions.

6.2   METHODS AND MATERIALS

6.2.1   HYPERTHERMIA PHASED ARRAY EQUIPMENT DESCRIPTION

The hyperthermia transmit phased array antenna system investigated in this chapter is a Model BSD-2000, SIGMA-60 applicator system developed by Turner [29, 30] (BSD Medical Corporation, Salt Lake City, Utah). By fully surrounding the patient with an annular phased array, it is possible to obtain constructive interference (or signal enhancement) deep within the target volume [31]. At the time of this research, clinical operation of the BSD-2000 hyperthermia phased array allowed manual control of the array transmit-element amplitude and phase. Some improvement in the electric-field distribution could be achieved by this manual trial-and-error method, but automatic adjustment techniques, offered by computer-controlled adaptive arrays, are desirable and possibly offer therapeutically better electric-field distributions. A candidate adaptive nulling algorithm is a gradient search based on minimizing the signal power received by electric-field sensors at the desired null positions, while maintaining the signal power at the focal point. The reason for considering this algorithm is that the system used in these experiments measured only the applied electric-field power and not the phase. Note that the E-field probes used are short dipoles with semiconductor diode detectors [32-34]. The concept of adaptive nulling as it applies to clinical hyperthermia will now be described.

Consider a potential clinical application of the adaptive phased array hyperthermia system. The four-channel hyperthermia system investigated here uses up to eight noninvasive electric-field probes to monitor clinical hyperthermia treatments. The eight E-field probes could provide feedback signals to the adaptive algorithm. In theory, with four transmit channels, three independent adaptive nulls can be formed by using any three of the eight feedback signals. There are various methods to select the desired null sites. All eight of the E-field probe sensors could be placed at desired positions, and then the adaptive algorithm could minimize the three largest measured electric-field signals on the patient's skin surface or in other areas away from the tumor site. If two or three positions are identified as potential hot spots, adaptive nulls could be automatically formed at those positions. Alternately, if during the heating therapy a patient can localize a painful tissue area, the nearest electric-field probe can be used as the feedback signal to the adaptive nulling algorithm to reduce the E-field at that location. Since the adaptive array transmit amplifiers and phase shifters could be continuously updated, patient breathing and patient motion should not affect the performance of the feedback algorithm. Also, the electrical conductivity and dielectric constant of tissue can vary as the tissue is heated. A variation in tissue electrical properties will produce a path-dependent phase shift, and the feedback algorithm can compensate for this phase shift. In some situations, more than three independent adaptive nulls may be needed to produce a therapeutic thermal distribution. An extension of the four-channel transmit phased array system to eight channels would provide up to seven independent adaptive nulls in normal tissue plus a desired focused beam directed at a tumor.

The hyperthermia phased array system used in these measurements has an approximate 60-cm array diameter with eight uniformly spaced dipole elements (dipole length is 44 cm) operating CW over the frequency band 60 to 120 MHz [29, 30]. The eight dipoles are fed as four active pairs of elements with up to 500W average CW power per channel. Each of the four active channels has an electronically controlled variable-phase shifter for phase focusing the array. The variable transmit amplifiers (0 to 500W) and variable phase shifters (0° to 180°) are controlled by digital-to-analog (D/A) converters. Temperature and electric-field probe sensors (both invasive and noninvasive) are used to monitor the treatment. A temperature-controlled cool-water bolus between the phased array and patient's skin surface is used to reduce skin surface temperatures during deep heating. The water bolus is filled with circulating deionized water, which has a very low RF propagation loss of about 0.0002 dB/cm at 100 MHz. The dielectric properties [35-37] of the water bolus and some of the other pertinent materials used in these experiments, with frequency range 100 to 120 MHz and room temperature (25°C), are summarized in Table 6.1.

Three types of electric-field probes are used in these measurements. For invasive electric-field measurements, a BSD Model EP-500 probe is used.

Table 6.1
Dielectric Properties in the Range of 100 to 120 MHz for Materials Used in the Experiments

Material	ϵ_r	σ (S/m)	λ (cm)	Loss (dB/cm)
Deionized water (bolus)	78.0	0.0001	33.9	0.0002
0.3% NaCl phantom	77.3	0.5	30.3	0.83
0.9% NaCl phantom	75.0	1.5	22.5	1.8
Plexiglass	2.7	0.00012	182.5	0.001
Polyethylene	2.25	0.0002	200	0
Polyvinyl chloride (PVC)	3.0	0.0005	173.1	0.005
Free space	1.0	0.0	300	0

For noninvasive electric-field measurements, BSD Model EP-100 and EP-400 probes are used. Each of the probes used in the measurements is axially oriented, with respect to the phased array hyperthermia applicator, and has high-impedance mechanically flexible transmission lines to reduce scattering effects from the electromagnetic field. The design of the EP-500 probe is as described by Astrahan [34]. The short metallic leads of a Schottky diode are connected between two highly resistive conductive leads. A 1-cm length of conducting wire is added to one of the resistive leads and responds to linear polarization. The leads of the diode and the additional conducting wire act as an offset-feed dipole antenna. The lengths of the dipole antennas used in the above three probes are as follows: EP-500 is a 1-cm dipole, EP-100 is a 3-cm dipole, and EP-400 is a 3-element array of dipoles with a total length of 23 cm. Two of the Bowman temperature probes provided with the hyperthermia system are used to monitor the beef phantom temperatures. In this chapter, assume that an EP-100 probe will provide localized adaptive nulls; thus, the EP-100 probe is used for providing the feedback signals. Note that the EP-400 field probes have also been used by the author for feedback signals and the measured null strengths are similar to the EP-100 measured data presented in this chapter.

6.2.2 SALINE-FILLED CYLINDRICAL PHANTOM DESCRIPTION

Photographs of the homogeneous saline-filled cylindrical phantom (28-cm diameter and 40-cm length), inside the hyperthermia phased array antenna applicator with the water bolus filled, are shown in Figure 6.2 (front view) and Figure 6.3 (side view). The cylindrical phantom [12, 14, 15] is a 23-liter polyethylene bottle filled with saline (0.9% NaCl) (9g/kg) having a dielectric constant of 75 and an electrical conductivity of 1.5 S/m, which has an attenuation of about 1.8 dB/cm as listed in Table 3.3. Three electric-field

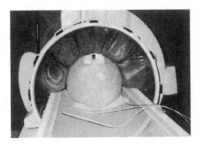

Figure 6.2 Front view of homogeneous saline-filled cylindrical phantom and electric-field probes located within a four-element dipole phased array hyperthermia system [12].

Figure 6.3 Side view of homogeneous saline-filled cylindrical phantom and electric-field probes located within a four-element dipole phased array hyperthermia system [12].

probes were used to monitor the electric-field distribution. One probe was mounted invasively at the center of the cylindrical phantom and two probes were mounted on the left and right sides of the phantom.

6.2.3 HETEROGENEOUS BEEF PHANTOM DESCRIPTION

The heterogeneous beef phantom and the probe positions used in these experiments are depicted in Figure 6.4. The beef phantom [12, 15] is a tapered cut obtained from the hind leg just above the knee. The front face of the beef has a horizontal width of 38 cm and a vertical height of 23 cm, providing dimensions comparable to the cross section of the human torso. The thickness of the beef is 15 cm and the weight is 10.9 kg. A photograph of the experimental setup for the beef phantom is shown in Figure 6.5. Three electric-field probes are used to monitor the electric-field distribution. One probe is mounted invasively at the center of the beef phantom and two probes are mounted on the left and right sides of the phantom. The power measured at the invasive probe site is taken as the power delivered to a fictitious tumor site.

6.2.4 LIGHT-EMITTING DIODE SALINE-FILLED ELLIPTICAL PHANTOM DESCRIPTION

For previous investigations that involved nonadaptive pretreatment planning for the deep-heating RF phased dipole ring array hyperthermia system, a light-emitting diode (LED) matrix phantom was constructed to display the effects of manually adjusting the amplitude and phase of the array antennas [38-41]. Figure 6.6 shows the LED matrix phantom [41] that consists of 137 LED sensors positioned in an elliptically shaped Plexiglass plate with a square-grid diode spacing of 2 cm. The LED matrix is positioned within an elliptical

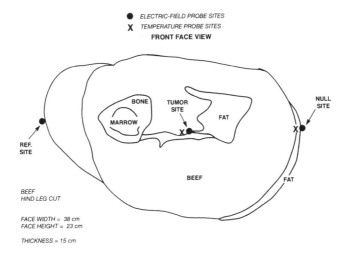

Figure 6.4 Front-face dimensions and probe positions for the heterogeneous beef phantom. (From [15] with permission from Informa Healthcare, www.informaworld.com.)

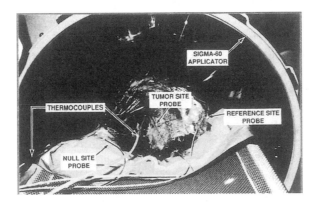

Figure 6.5 Beef phantom target with electric-field probes and thermocouples [12].

cylinder of homogeneous saline solution contained within a 2-mm thick PVC shell with dielectric parameters as listed in Table 6.1. The saline concentration used here is 0.3% NaCl (3 g/kg) which results in a dielectric constant of 77.3 and an electrical conductivity of 0.5 S/m (refer to Table 6.1) – these dielectric parameters are similar to many of the tissues listed in Table 3.3. The cylinder has a cross section of 24 cm × 36 cm, similar to the cross section of a human torso, and has transparent ends for viewing the LEDs. The length of the LED metallic leads forming the dipole receive sensor is nominally 5 cm tip-to-tip. The light output from the LEDs is directly proportional to

Figure 6.6 Light-emitting diode (LED) matrix phantom, which simulates a cross section of a human torso, is used in treatment planning for clinical hyperthermia treatments. The photograph shows the LED dipole-sensor array removed from the inside of the elliptical phantom shell [19].

the local electric-field strength generated by the RF ring array. This phantom has been used for pretreatment planning of clinical hyperthermia trials in the following manner. With the matrix phantom load centered in the RF ring array aperture, the operator would begin by manually adjusting the RF power amplifiers and phase shifters until the LED phantom visually demonstrated maximum electric-field strength in the planned tumor target with as few hot spots as possible in healthy tissue. These manual adjustments of amplitude and phase are time consuming and can take more than two hours to complete. Then, the patient would be substituted for the phantom and the clinical treatment would be conducted – the assumption being that the irradiation pattern would not change substantially after substituting the patient for the phantom. Pretreatment planning with this manual trial-and-error adjustment procedure often produces unacceptable RF hot spots in healthy tissue for the required deep-tumor heating [39]. Thus, the experiments here investigated pretreatment planning with the LED phantom [19-21] using computer-controlled adaptive nulling [12, 14, 15, 19-22] for eliminating the unwanted hot spots and focusing RF energy in the tumor. Figure 6.7 shows the electric-field probes attached to the LED phantom, and Figure 6.8 shows the LED phantom located within the dipole phased array hyperthermia system.

To localize energy deposition for an appropriate temperature rise in a deep tumor, it is necessary first to monitor the electric-field magnitude $|E|$ received at one or more feedback probes located within the tumor and adjacent healthy tissues, and then adjust the amplitude and phase of each transmitting antenna of the array for maximum RF power deposition within the tumor and minimum power deposition in healthy tissue. A gradient-search computer algorithm (see Chapter 2) that modifies antenna array input parameters (drive signals) based on the rate of change of system output parameters (power

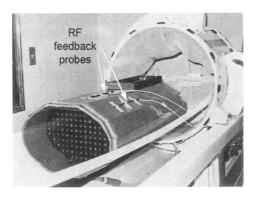

Figure 6.7 Light-emitting diode (LED) matrix phantom with electric-field probes attached to the surface [19].

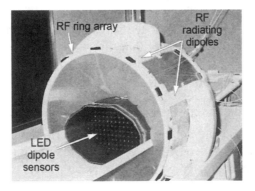

Figure 6.8 Light-emitting diode (LED) matrix phantom located within the dipole phased array hyperthermia applicator [19].

deposition pattern) can be used to determine adaptively the individual antenna power and phase input signals to maximize (focus) or minimize (null) the electromagnetic radiation measured at one or more feedback probe positions [9, 10, 12, 14, 15, 20-22]. The adaptive nulling approach used is based on algorithms developed for adaptive phased array radar and communications signal processing systems [2]. The minimum resolution width of an adaptive null is approximately equal to the half-power radiation beam width of the adaptive array antenna. This property allows an adaptive null formed on the surface of the body to reduce the electric field in regions such as healthy tissue that extend to some depth below the null. For hyperthermia applications, adaptive array nulls that reduce the RF power deposition by approximately 50% are assumed by the author to be effective in reducing or eliminating undesired hot spots in healthy tissues.

A computer program that implements the adaptive array gradient-search

Table 6.2
List of Adaptive Phased Array Hyperthermia Experiments

Expt. No.	Phantom	Frequency	Feedback
1	Saline-filled cylinder	100 MHz	2 Nulls
2	Saline-filled cylinder	100 MHz	1 Focus
3	Heterogeneous beef	120 MHz	1 Null
4	LED saline-filled elliptical	100 MHz	3 Nulls (anterior)
5	LED saline-filled elliptical	100 MHz	3 Nulls (lateral)

algorithm was developed by the author and integrated, in collaboration with G.A. King at the State University of New York (SUNY) Health Science Center in Syracuse, for phantom testing with the BSD-2000 hyperthermia system amplifier/phase shifter control and electric-field monitoring software at SUNY [12, 14, 15]. The integrated software implemented either adaptive nulling or adaptive focusing algorithms and monitored the output power of the electric-field probes during the gradient search. Similar software was implemented for phantom testing on the BSD-2000 hyperthermia system in collaboration with V. Sathiaseelan at Northwestern Memorial Hospital (NMH) in Chicago, Illinois [19-21]. In these experiments, the adaptive nulling algorithm used phase and amplitude control to form the nulls—the adaptive focusing algorithm used only phase control to maximize the focal-point electric field while maintaining constant transmit power.

6.3 MEASURED RESULTS

In this section, measured data are presented for a saline-filled cylindrical phantom, a heterogeneous beef phantom, and a light-emitting diode saline-filled phantom as summarized in Table 6.2. The measurement frequencies were 100 MHz for the homogeneous saline-filled cylindrical phantom and the LED saline-filled elliptical phantom and 120 MHz for the heterogeneous beef phantom. These frequencies were selected by monitoring the reflected power level of the transmitter channels as a function of frequency. Frequencies where the reflected power is low were determined by trial and error.

In the first experiment (Table 6.2), two adaptive nulls were formed independently on the left and right sides of the saline-filled cylindrical phantom while monitoring the deep tumor-site power. The longitudinal characteristics of one of the adaptive nulls were quantified by measurement. In the second experiment, an adaptive focusing gradient search was run for the cylindrical phantom with the focus positioned on the surface of the phantom. In the third experiment, one adaptive null is formed on the surface of the beef phantom. Temperature data are presented for this case. In the fourth experiment, three anterior adaptive nulls were formed for the LED saline-filled elliptical phantom. In the fifth experiment, three lateral adaptive nulls were formed for the LED saline-filled elliptical phantom.

In the following sections, the measured electric-field probe A/D converter data of the hyperthermia system data recording system is converted to power in decibels (dB) by computing $10 \log_{10}(p)$ where p is the value (measured power) sampled by the A/D converter.

6.3.1 SALINE-FILLED CYLINDRICAL PHANTOM: TWO ADAPTIVE NULLS

The first experiment [12, 14, 15] involves simultaneously forming two adaptive transmit nulls at two independent electric-field probe feedback positions. The invasive EP-500 probe was mounted in the center of the cylindrical phantom and measured the electric-field power at the simulated tumor site. Note that the EP-500 probe was inserted in a catheter through a rubber plug on top of the phantom. Two EP-100 probes were attached to the left and right sides of the outer surface of the phantom and measured the adaptive-nulling feedback signals.

A 50-iteration adaptive nulling gradient search (without acceleration) was executed at 100 MHz for this test configuration. The gradient-search step sizes in terms of D/A converter states are $\Delta A = 10$ and $\Delta \phi = 50$. The D/A converters that control the transmit amplifiers and phase shifters each have 12 bits, thus 4096 states. The commanded initial transmit forward powers on channels 1 to 4 are 215W, 215W, 215W, and 215W, respectively (860W total array power). The numbering of transmit-array elements 1, 2, 3, and 4 correspond to the anterior, posterior, left, and right quadrants, respectively. The right side refers to the side of the phased array applicator that contains the array's coaxial feed cables. The commanded initial phase shifts are set to mid-range values of 89°, 89°, 89°, and 89° in quadrants 1 to 4, respectively. After 50 iterations of the adaptive nulling algorithm, the transmit commanded forward powers are 510W, 55W, 190W, and 76W (for a total array power of 831W) and the corresponding phase shifts are 0°, 99°, 58°, and 101°. The measured electric field as a function of iteration number is displayed in Figure 6.9. There are three curves plotted; each curve is normalized to 0 dB at the initial iteration. The adaptive cancellation is defined in decibels as $10 \log_{10}$ (power after nulling divided by power before nulling). The two null-site probe powers decrease nearly monotonically over the 50 iterations. At iteration 50, the cancellation for null-site probe #2 is 18.0 dB, while the cancellation for null-site probe #3 is 11.5 dB. The tumor-site probe power

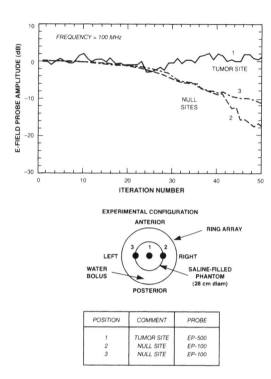

Figure 6.9 Experimental demonstration of an adaptive nulling hyperthermia system. The figure shows the measured electric field as a function of adaptive nulling gradient-search iteration number for a saline-filled cylindrical phantom. Two adaptive nulls are independently formed at electric-field probe positions 2 and 3. (From [15] with permission from Informa Healthcare, www.informaworld.com.)

stays relatively constant during the 50 iterations. With the two symmetrically formed nulls, the main beam tends to stay centered at the desired focus. Note that the orientation of the invasive probe in a patient cannot always be axially directed as with these phantom measurements. For deep-seated probes in a patient, a radial probe orientation is more readily achieved. Thus, monitoring the power delivered to a deep-seated tumor site can be difficult. However, for this experiment the invasive probe measured data is not used in the adaptive nulling array feedback algorithm, and the observation of the power staying relatively constant at the tumor site should hold.

Using the above adaptive weights from iteration number 50, a measurement of the electric-field amplitude in the vicinity of one of the nulls was performed. Probe #3 was moved uniformly to nine positions longitudinally on the surface of the cylindrical phantom; the nine positions scanned 20 cm.

The electric field before and after nulling is presented in Figure 6.10. Before adaptive nulling, the radiation pattern has broad coverage as expected in the near field of a dipole antenna, and the half-power beamwidth is greater than 20 cm. After adaptive nulling, the electric-field magnitude is reduced at each measurement position over the 20-cm longitudinal region. The electric field tends to increase as the probe moves away from the null position at $y = 0$.

6.3.2 SALINE-FILLED CYLINDRICAL PHANTOM: ADAPTIVE FOCUS

The second experiment [12, 14, 15] involves the formation of an adaptive focus (power maximum) on the surface of the cylindrical phantom. Here, the experiment simply demonstrates that the adaptive transmit array algorithm can increase the power delivered to a selected position. One EP-100 probe

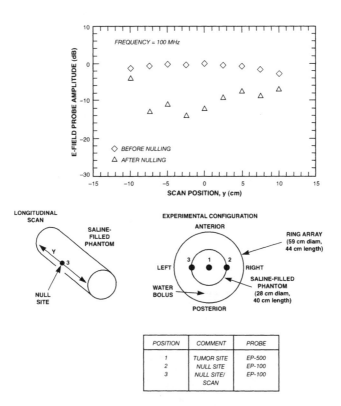

Figure 6.10 Measured electric field in a longitudinal scan on the surface of the cylindrical phantom before and after adaptive nulling. (From [15] with permission from Informa Healthcare, www.informaworld.com.)

was taped to the right side of the phantom and measured the adaptive focusing feedback signal. A 30-iteration adaptive-focusing gradient search was executed for this test configuration at 100 MHz. The gradient-search step sizes in terms of raw D/A converter states are $\Delta A = 0$ and $\Delta \phi = 50$. Because the amplitude step size is zero, only phase focusing is in effect. The initial transmit forward powers on channels 1 to 4 are 215W, 215W, 215W, and 215W and the corresponding transmit phase shifts of the four channels are 100°, 100°, 100°, and 100°. Since the transmit phase shifts are equal and the homogeneous target geometry is circularly symmetric, the initial focus is expected to occur near the center of the saline phantom. After 30 iterations of the phase-focusing algorithm, the transmit forward powers are held constant at 215W and the phase shifts are 13°, 98°, 78°, and 30°. The measured electric field as a function of iteration number is displayed in Figure 6.11. The focal-probe power increases by approximately 1 dB over 30 iterations. The data indicate convergence in about 15 iterations. The experiment did not attempt to measure the radiation pattern for this case. Such a measurement is necessary to demonstrate that the phase focusing has formed a beam with peak radiation at the desired surface focal point.

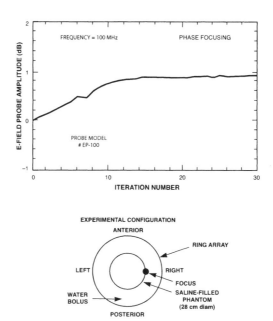

Figure 6.11 Measured electric field as a function of adaptive phase focusing gradient-search iteration number for a saline-filled cylindrical phantom. (From [15] with permission from Informa Healthcare, www.informaworld.com.)

6.3.3 ONE ADAPTIVE NULL FOR HETEROGENEOUS BEEF PHANTOM

The third experiment [12, 15] demonstrates the formation of one adaptive null for the heterogeneous beef phantom. The invasive EP-500 probe is mounted in the center of the beef phantom and measures the electric-field power at the simulated tumor site. An EP-400 probe is taped to the left side of the phantom and an EP-100 probe is taped to the right side of the phantom. The EP-400 probe measures a reference electric field, while the EP-100 probe measures the adaptive nulling feedback signal. A 50-iteration adaptive nulling gradient search was executed at 120 MHz for this test configuration. The gradient-search step sizes in terms of raw D/A converter states are $\Delta A = 10$ and $\Delta \phi = 50$. The initial transmit forward powers on channels 1 to 4 are 215W, 215W, 215W, and 215W and the corresponding phase shifts are $103°$, $103°$, $103°$, and $103°$. After 50 iterations of the nulling algorithm, the transmit forward powers are 163W, 390W, 170W, and 112W (total array power equals 845W) and the transmit phase shifts are $76°$, $60°$, $58°$, and $131°$. The measured electric field as a function of iteration number is displayed in Figure 6.12. The null-site probe power decreases during the first 30 iterations. After 30 iterations, the average null-site amplitude (cancellation) is about 20 dB. The tumor-site power has dropped by about 2 dB at iteration 50; this power reduction is attributed in part to the beam peak shifting away from the central focus. The reference site power is observed to increase by about 2 dB during the 50 iterations, which is consistent with the beam peak shift assumption.

To record calibrated temperature data, the RF power was turned off (for about five minutes) after each set of 10 iterations. During the period when the RF power is off, the temperature-monitoring mode of the BSD-2000 hyperthermia system is accessed and the temperature data recorded. The RF is then turned back on with the transmit weights set to the previous iteration. The measured temperatures at the tumor site and null site are shown in Figure 6.13. The initial temperatures were approximately equal; after 80 minutes there is a $4°C$ higher temperature at the tumor site compared to the null site. Thus, the data suggest that adaptive nulling may be effective in improving the thermal distribution in hyperthermia.

6.3.4 THREE ANTERIOR ADAPTIVE NULLS FOR LED SALINE-FILLED ELLIPTICAL PHANTOM

For the fourth experiment [19-21], three anterior adaptive nulls were formed on the LED saline-filled elliptical phantom. A radiofrequency of 100 MHz was used for the ring array system, and the sum of the input power to all four channels was held constant at 860W. An invasive catheter with a dipole

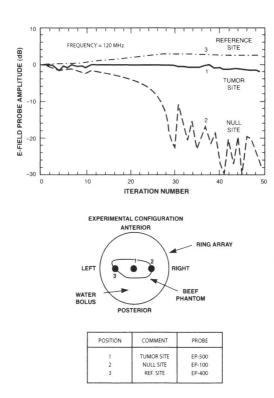

Figure 6.12 Measured electric field as a function of adaptive nulling gradient-search iteration number for a beef phantom. One adaptive null is formed at probe position 2. (From [15] with permission from Informa Healthcare, www.informaworld.com.)

electric-field sensor was positioned at a depth of approximately 8 cm in the lower half of the phantom and used to measure the local electric field at the simulated deep-seated tumor site. Three independent noninvasive electric-field probes, spaced circumferentially at 10-cm intervals, were attached to the upper-half surface of the phantom (refer to Figure 6.7) and used to measure feedback signals for reducing local power deposition on the upper (anterior) surface. The goal of the experiment was to irradiate only the lower portion of the phantom, which represented the location of a fictitious tumor (such as rectal cancer), while minimizing irradiation of the upper portion that contained a region of simulated healthy tissue.

The power and phase input to each of the four RF radiating antennas of the ring array were manually set initially to equal values; that is, $P_1 = P_2 = P_3 = P_4 = 215\text{W}$ and $\phi_1 = \phi_2 = \phi_3 = \phi_4 = 89°$. The computer started the adaptive array algorithm by automatically adjusting, via digital-to-analog

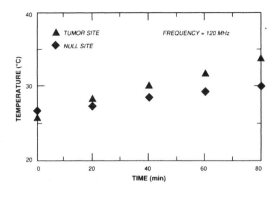

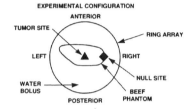

Figure 6.13 Measured temperatures in the beef phantom heated by the adaptive phased array. Eighty minutes after the start of the experiment, a 4°C higher temperature exists at the tumor site compared to the null site. (From [15] with permission from Informa Healthcare, www.informaworld.com.)

converters, the power amplifiers and phase shifters in each of the four channels of the phased array. The computer software performed calculations of the rate of change of the measured RF power at the surface sensors (simulated healthy tissue regions) after each adjustment of RF power and phase for the array transmit channels. For this experiment, the method of steepest descent algorithm was used to determine the input power and phase commands that minimize the summation of the local power deposition measured by each surface E-field feedback sensor. All adjustments are completed and the adaptive nulls formed in approximately 2 minutes, which is an appropriate speed for real-time use in optimizing clinical treatments. Once the adaptive nulls are formed, they can be maintained throughout the treatment.

Before adaptive nulling, both the electric-field sensors and the LED matrix phantom indicated multiple hot spots as shown in Figures 6.14 and 6.15, respectively. In Figure 6.15, since the desired heating zone is posterior, the hot spots are considered to be in the anterior portion of the phantom. Then, the adaptive nulling algorithm was executed until it reduced the RF

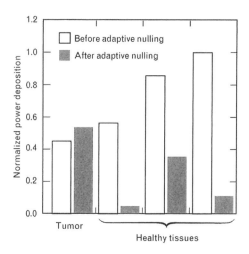

Figure 6.14 Measured electric-field probe data in a pretreatment planning LED saline-filled elliptical phantom, before and after adaptive nulling at three anterior surface sites [19].

feedback signal at each surface nulling sensor by at least a factor of two. The electric-field sensors and the LED matrix phantom then displayed the electric-field distributions after adaptive nulling, as shown in Figures 6.14 and 6.16. In Figure 6.14, the measured E-field probe data for the elliptical treatment-planning phantom demonstrates the effect of adaptive nulling at three independent surface sites to protect healthy tissue regions while an 8-cm deep tumor site is irradiated with a coherent four-channel ring phased array system operating at a radiofrequency of 100 MHz. The normalized RF power deposition at each E-field sensor before adaptive nulling is indicated by the white bars. Prior to adaptive nulling, the RF power deposition in healthy tissues is greater than the RF power deposition in the tumor. The normalized RF power deposition after nulling (shaded bars) measured at the simulated deep tumor site increases by 19%, while the RF power deposition measured by the three E-field feedback probes on the surface of the phantom is reduced by 91%, 57%, and 87%, respectively. The LED display in Figure 6.16 shows that the simulated tumor position in the lower half of the phantom is fully irradiated while the upper half of the phantom, containing the region of simulated healthy tissue, has a substantially reduced electric-field intensity. The measurements demonstrate that the adaptive nulling process results in a stronger irradiation of the tumor region compared to the irradiation of the superficial healthy tissues.

Figure 6.15 Pretreatment planning LED matrix phantom irradiated by an adaptively controlled coherent RF ring phased array at 100 MHz, before adaptive nulling [19].

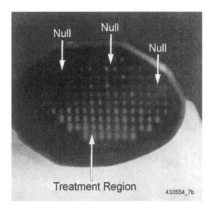

Figure 6.16 Pretreatment planning LED matrix phantom irradiated by an adaptively controlled coherent RF ring phased array at 100 MHz, after adaptive nulling at three anterior surface sites [19].

6.3.5 THREE LATERAL ADAPTIVE NULLS FOR LED SALINE-FILLED ELLIPTICAL PHANTOM

For the fifth experiment [19-21], three adaptive nulls were formed in/on the LED saline-filled elliptical phantom. A radiofrequency of 100 MHz was used for the ring array system, and the sum of the input power to all four channels was held constant at 860W. An invasive catheter with a dipole electric-field sensor was positioned at a depth of approximately 8 cm in the lower left portion of the phantom and used to measure the local electric field at the simulated deep-seated tumor site. Three independent noninvasive electric-field probes, spaced circumferentially at 10-cm intervals, were attached to

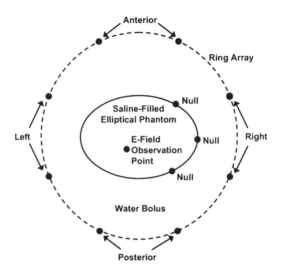

Figure 6.17 Sketch of geometry for hyperthermia phased array, three lateral surface nulling sensors, and light-emitting diode (LED) matrix phantom for adaptive nulling experiments.

the right-half surface of the phantom (refer to Figure 6.17) and were used to measure feedback signals for reducing local power deposition on the right side of the phantom. The goal of the experiment was to irradiate only the left portion of the phantom, which represented the location of a ficticious tumor, while minimizing irradiation of the right portion, which contained a region of simulated healthy tissue.

The power and phase input to each of the four RF radiating antennas of the ring array were manually set initially to equal values, that is, $P_1 = P_2 = P_3 = P_4 = 215$W and $\phi_1 = \phi_2 = \phi_3 = \phi_4 = 89°$. The computer started the adaptive array algorithm by automatically adjusting, via digital-to-analog converters, the power amplifiers and phase shifters in each of the four channels of the phased array. The computer software performed calculations of the rate of change of the measured RF power at the surface sensors (simulated healthy tissue regions) after each adjustment of RF power and phase to the array transmit channels. For this experiment, the method of steepest descent algorithm was used and all adjustments are completed and the adaptive nulls formed in approximately 2 minutes.

Before adaptive nulling, both the electric-field sensors and the LED matrix phantom indicated multiple hot spots as shown in Figures 6.18 and 6.19. The adaptive nulling algorithm was executed until it reduced the RF feedback signal at each surface nulling sensor. The electric-field sensors and

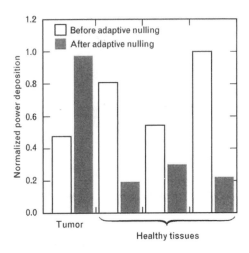

Figure 6.18 Measured electric-field probe data in a pretreatment planning LED saline-filled elliptical phantom, before and after adaptive nulling at three lateral surface sites (right side) [19].

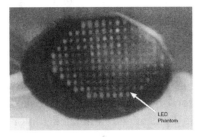

Figure 6.19 Pretreatment planning LED matrix phantom irradiated by an adaptively controlled coherent RF ring phased array at 100 MHz before adaptive nulling on the right lateral side [19].

the LED matrix phantom then displayed the electric-field distributions after adaptive nulling as shown in Figures 6.18 and 6.20. In Figure 6.18, the measured E-field probe data for the elliptical treatment-planning phantom demonstrate the effect of adaptive nulling at three independent surface sites to protect healthy tissue regions while an 8-cm deep tumor site is irradiated with a coherent four-channel ring phased array system operating at a radiofrequency of 100 MHz. The normalized RF power deposition at each E-field sensor before adaptive nulling is indicated by the white bars. Prior to adaptive nulling, the RF power deposition in healthy tissues is greater than the RF power deposition in the tumor. The normalized RF power deposition after nulling (shaded bars) measured at the simulated deep tumor

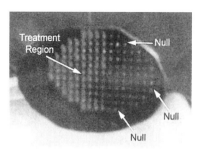

Figure 6.20 Pretreatment planning LED matrix phantom irradiated by an adaptively controlled coherent RF ring array at 100 MHz after adaptive nulling on the right lateral side of the phantom [19].

site increases by 3.1 dB, while the RF power deposition measured by the three E-field feedback probes on the surface of the phantom is reduced by 6.4, 2.6, and 6.5 dB, respectively. The LED display in Figure 6.20 shows that the simulated tumor position in the left half of the phantom is irradiated while the right half of the phantom, containing the region of simulated healthy tissue, has a substantially reduced electric-field intensity. The measurements demonstrate that the adaptive nulling process results in a stronger irradiation of the tumor compared to the irradiation of the superficial healthy tissues. Finally, Figure 6.21 shows the normalized measured electric field at the three adaptive null sites and at the simulated tumor site as a function of iteration index. After only one iteration, the measured data show a substantial increase (2.7 dB) in the electric field at the simulated tumor site and a substantial decrease (3.7, 1.6, and 5.0 dB) in the electric field at the null sites.

6.4 SUMMARY

In this chapter the adaptive phased array gradient-search algorithm described in Chapter 2 was evaluated in three types of phantom target: a saline-filled cylindrical phantom, a heterogeneous beef phantom, and a light-emitting diode saline-filled elliptical phantom. For these tests, a dipole phased array hyperthermia system was used. The initial tests in the homogeneous saline phantom demonstrated both adaptive nulling and adaptive focusing capabilities. The beef phantom tests demonstrated adaptive nulling in a heterogeneous body. The tests in the light-emitting diode saline-filled elliptical phantom demonstrated that a region of simulated tissue can be irradiated while nulling the electric field on the body surface in a noninvasive manner. The next chapter describes a monopole phased array hyperthermia

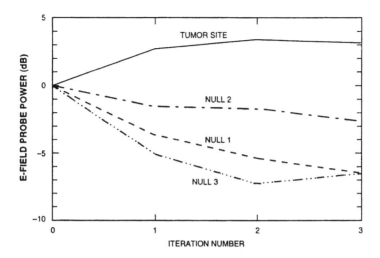

Figure 6.21 Measured electric-field probe power versus iteration, for a pretreatment planning LED matrix phantom irradiated by an adaptively controlled coherent RF ring array at 100 MHz, during adaptive nulling on the right lateral side of the phantom [19].

applicator that could potentially be used in treating deep-seated cancers with the noninvasive adaptive phased array thermotherapy technique.

References

[1] Fenn, A.J., et al., "The Development of Phased Array Radar Technology," *Lincoln Laboratory Journal*, Vol. 12, No. 2, 2000, pp. 321-340.

[2] Fenn, A.J., *Adaptive Antennas and Phased Arrays for Radar and Communications,* Norwood, MA: Artech House, 2008.

[3] Fenn, A.J., "Theory and Analysis of Near Field Adaptive Nulling," *1986 IEEE Antennas and Propagation Symposium Digest,* Vol. 2, New York: 1986, pp. 579-582.

[4] Fenn, A.J., "Theory and Analysis of Near Field Adaptive Nulling," *Proc Asilomar Conf Signals, Systems and Computers,* Computer Society Press of the IEEE, Washington, D.C.: Nov 10-12, 1986, pp. 105-109.

[5] Fenn, A.J., "Evaluation of Adaptive Phased Array Far-Field Nulling Performance in the Near-Field Region," *IEEE Trans. Antennas Propagat.*, Vol. 38, No. 2, 1990, pp. 173-185.

[6] Fenn, A.J., H.M. Aumann, F.G. Willwerth, and J.R. Johnson, "Focused Near-Field Adaptive Nulling: Experimental Investigation," *1990 IEEE Antennas Propagation Soc. Int. Symp. Digest,* Vol. 1, May 7-11, 1990, pp. 186-189.

[7]   Fenn, A.J., "Analysis of Phase-Focused Near-Field Testing for Multiphase-Center Adaptive Radar Sysems," *IEEE Trans. Antennas Propagat.*, Vol. 40, No. 8, 1992, pp. 878-887.

[8]   Fenn, A.J., "Moment Method Analysis of Near Field Adaptive Nulling," *IEE Sixth Int. Conf. on Antennas and Propagation, ICAP 89*, April 4-7, 1989, pp. 295-301.

[9]   Fenn, A.J., "Adaptive Nulling Hyperthermia Array," US Patent No. 5,251,645, October 12, 1993.

[10] Fenn, A.J., "Adaptive Focusing and Nulling Hyperthermia Annular and Monopole Phased Array Applicators," US Patent No. 5,441,532, August 15, 1995.

[11] Fenn, A.J., "Non-Invasive Adaptive Nulling for Improved Hyperthermia Thermal Dose Distribution," *IEEE Engineering in Medicine and Biology Society Int Conf*, October 31 - November 3, 1991, Vol. 13, No. 2, pp. 976-977.

[12] Fenn, A.J., and G.A. King, "Adaptive Nulling in the Hyperthermia Treatment of Cancer," *The Lincoln Laboratory Journal*, Lincoln Laboratory, Massachusetts Institute of Technology, Vol. 5, No. 2, 1992, pp. 223-240.

[13] Fenn, A.J., C.J. Diederich, and P.R. Stauffer, "An Adaptive-Focusing Algorithm for a Microwave Planar Phased-Array Hyperthermia System," *The Lincoln Laboratory Journal*, Lincoln Laboratory, Massachusetts Institute of Technology, Vol. 6, No. 2, 1993, pp. 269-288.

[14] Fenn, A.J., and G.A. King, "Experimental Investigation of an Adaptive Feedback Algorithm for Hot Spot Reduction in Radio-Frequency Phased-Array Hyperthermia," *IEEE Trans Biomed Eng.*, Vol. 43, No. 3, 1994, pp. 273-280.

[15] Fenn, A.J., and G.A. King, "Adaptive Radio Frequency Hyperthermia Phased Array System for Improved Cancer Therapy: Phantom Target Measurements," *Int J Hyperthermia*, Vol. 10, No. 2, 1994, pp. 189-208.

[16] Fenn, A.J., B.A. Bornstein, G.K. Svensson, and H.F. Bowman, "Minimally Invasive Monopole Phased Arrays for Hyperthermia Treatment of Breast Carcinomas: Design and Phantom Tests," *Int. Symp. on Electromagnetic Compatibility*, Sendai, Japan: Vol. 10, No. 2, 1994, pp. 566-569.

[17] Fenn, A.J., "Minimally Invasive Monopole Phased Arrays for Hyperthermia Treatment of Breast Cancer," In: *Proc. 1994 Int. Symp. on Antennas*, Nice, France: November 8-10, 1994 418-421.

[18] Fenn, A.J., "Minimally Invasive Monopole Phased Array Hyperthermia Applicators and Method for Treating Breast Carcinomas," US Patent No. 5,540,737, July 30, 1996.

[19] Fenn, A.J., V. Sathiaseelan, G.A. King, and P.R. Stauffer, "Improved Localization of Energy Deposition in Adaptive Phased-Array Hyperthermia Treatment of Cancer," *The Lincoln Laboratory Journal*, Lincoln Laboratory, Massachusetts Institute of Technology, Vol. 9, No. 2, 1996, pp. 187-196.

[20] Sathiaseelan, V., A.J. Fenn, and A. Taflove, "Recent Advances in External Electromagnetic Hyperthermia," In: Chapter 10 of *Advances in Radiation Treatment*, Mittal, B.B., J.A. Purdy, and K.K. Ang, (eds.), Boston, Massachusetts: Kluwer Academic Publishers, 1998, pp. 213-245.

[21] Fenn, A.J., V. Sathiaseelan, G.A. King, and P.R. Stauffer, "Improved Localization of Energy Deposition in Adaptive Phased Array Hyperthermia Treatment of Cancer," *J Oncol Management*, Vol. 7, No. 2, 1998, pp. 22-29.

[22] Fenn, A.J., "Thermodynamic Adaptive Phased Array System for Activating Thermosensitive Liposomes in Targeted Drug Delivery," US Patent No. 5,810,888, September 22, 1998.

[23] Fenn, A.J., G.L. Wolf, and R.M. Fogle, "An Adaptive Phased Array for Targeted Heating of Deep Tumors in Intact Breast: Animal Study Results," *Int J Hyperthermia*, Vol. 15, No. 1, 1999, pp. 45-61.

[24] Gavrilov, L.R., J.W. Hand, J.W. Hopewell, and A.J. Fenn, "Pre-clinical Evaluation of a Two-Channel Microwave Hyperthermia System with Adaptive Phase Control in a Large Animal," *Int J Hyperthermia*, Vol. 15, No. 6, 1999, pp. 495-507.

[25] Gardner, R.A., H.I. Vargas, J.B. Block, C.L. Vogel, A.J. Fenn, G.V. Kuehl, and M. Doval, "Focused Microwave Phased Array Thermotherapy for Primary Breast Cancer," *Ann Surg Oncol*, Vol. 9, No. 4, 2002, pp. 326-332.

[26] Vargas, H.I., W.C. Dooley, R.A. Gardner, K.D. Gonzalez, S.H. Heywang-Kobrunner, and A.J. Fenn, "Focused Microwave Phased Array Thermotherapy for Ablation of Early-Stage Breast Cancer: Results of Thermal Dose Escalation," *Ann Surg Oncol*, Vol. 11, No. 2, 2004, p. 139-146.

[27] Fenn, A.J., *Breast Cancer Treatment by Focused Microwave Thermotherapy*, Sudbury, MA: Jones and Bartlett, 2007, pp. 54-56.

[28] Vargas, H.I., W.C. Dooley, A.J. Fenn, M.B. Tomaselli, and J.K. Harness, "Study of Preoperative Focused Microwave Phased Array Thermotherapy in Combination With Neoadjuvant Anthracycline-Based Chemotherapy for Large Breast Carcinomas," *Cancer Therapy*, Vol. 5, 2007, pp. 401-408, published online (www.cancer-therapy.org), November 25, 2007.

[29] Turner, P.F., T. Schaefermeyer, and T. Saxton, "Future Trends in Heating Technology of Deep-Seated Tumors," *Recent Results in Cancer Research*, Vol. 107, 1988, pp. 249-262.

[30] Turner, P.F., A. Tumeh, and T. Schaefermeyer, "BSD-2000 Approach for Deep Local and Regional Hyperthermia: Physics and Technology," *Strahlentherapie Onkologie*, Vol. 165, No. 10, 1989, pp. 738-741.

[31] Sullivan, D., "Mathematical Methods for Treatment Planning in Deep Regional Hyperthermia," *IEEE Trans. Microwave Theory and Techniques*, Vol. 39, No. 5, 1991, pp. 864-872.

[32] Bassen, H.I., and G.S. Smith, "Electric Field Probes–A Review," *IEEE Trans on Antennas and Propagation*, Vol. AP-31, No. 5, 1983, pp. 1710-1718.

[33] Uhlir, A., "Characterization of Crystal Diodes for Low-Level Microwave Detection," *The Microwave Journal*, July 1963, pp. 59-67.

[34] Astrahan, M., et.al., "Heating Characteristics of a Helical Microwave Applicator for Transurethral Hyperthermia of Benign Prostatic Hyperplasia," *Int. J. of Hyperthermia*, Vol. 7, No. 1, 1991, pp. 141-155.

[35] von Hippel, A.R., (ed.), *Dielectric Materials and Applications,* New York: John Wiley, 1954.

[36] Stogryn, A., "Equations for Calculating the Dielectric Constant of Saline Water," *IEEE Trans Microwave Theory and Techniques,* Vol. 19, No. 8, 1971, pp. 733-736.

[37] Gabriel, S., R.W. Lau, and C. Gabriel, "The Dielectric Properties of Biological Tissues: Part III. Parametric Models for the Dielectric Spectrum of Tissues," *Phys Med Biol,* Vol. 41, 1996, pp. 2271-2293.

[38] Wust, P., et al., "Quality Control of the SIGMA Applicator Using a Lamp Phantom: a Four-Centre Comparison," *Int. J. Hyperthermia,* Vol. 11, No. 6, 1995, pp. 755-768.

[39] Mittal, B.B., et al., "Regional Hyperthermia in Patients with Advanced Malignant Tumors: Experience with the BSD 2000 Annular Phased-Array System and Sigma-60 Applicator," *Endocurietherapy/Hyperthermia Oncology,* Vol. 10, 1994, pp. 223-236.

[40] Schneider, C.J., and J.D.P. van Dijk, "Visualization by a Matrix of Light-Emitting Diodes of Interference Effects from a Radiative Four-Applicator Hyperthermia System," *Int. J. of Hyperthermia,* Vol. 7, No. 2, 1991, pp. 355-366.

[41] Schneider, C.J., J.D.P. van Dijk, A.A.C. De Leeuw, P. Wust and W. Baumhoer, "Quality Assurance in Various Radiative Hyperthermia Systems Applying a Phantom with LED Matrix," *Int. J. of Hyperthermia,* Vol. 10, No. 5, 1994, pp 733-747.

7

Monopole Phased Array for Deep Cancer

7.1 INTRODUCTION

For deep tumor heating, radiofrequency phased array hyperthermia applicators have received considerable attention in the literature for preclinical and clinical studies [1-30]. Adaptive phased array thermotherapy applicators have been explored by the author and colleagues [32-52]. As described in Chapter 1 (see Figure 1.8), a phased array hyperthermia system composed of four waveguides arranged in a ring with a coupling bolus for deep tumor heating was first introduced conceptually by von Hippel in 1973 [31]. A radiofrequency dipole ring phased array with a large water bolus, for deep tumor heating, has been developed by Turner [13, 14] and has been used in numerous preclinical and clinical studies [17-19, 21, 26-28], and in adaptive phased array preclinical studies by the author and colleagues [35, 37, 38, 42-44]. One of the difficulties in treating patients with a large-diameter hyperthermia array without a waveguide enclosure is the requirement for a large water bolus to couple the RF energy in toward the body. The mass of the large water bolus resting on the patient's body tends to be uncomfortable for the patient. An elliptical array applicator with a smaller water bolus can be used to reduce the mass of the water bolus resting on the patient [29]. An alternate approach to reducing the mass of the water bolus resting on the patient is to use a monopole phased array waveguide applicator as developed by the author [41, 45, 46], which is described and analyzed in this chapter.

7.2 METHODS AND MATERIALS

The previous chapter described adaptive nulling and focusing experiments with a dipole phased array hyperthermia system. A previous monopole array that was developed for 915-MHz thermotherapy treatment of breast cancer was investigated in [39, 40]. The present chapter describes the design and analysis of a monopole phased array applicator for deep heating, which could be used for preclinical and clinical adaptive phased array thermotherapy testing. A pictorial view of a monopole ring array applicator for thermotherapy is depicted in Figure 7.1. In this diagram there are eight monopole antenna elements mounted in the interior portion of a water-filled metallic waveguide cavity. The metallic waveguide cavity is designed so that the monopole array elements form a ring about a treatment aperture. In that way, each of the radiating monopoles can contribute to the heating of a deep tumor. For deep heating, the desired radiating frequency for this monopole phased array applicator is in the range of about 80 to 150 MHz. At 100 MHz, in free space the wavelength is 300 cm. The dielectric constant of deionized water at 100 MHz is approximately 78.0, and the electrical conductivity is approximately 0.0001 S/m [53, 54], and the wavelength is computed from (3.35) and (3.36) to be about 34 cm.

As described here, the array of monopole antenna elements is circular shaped and can have a diameter in the range of about 50 to 90 cm. The metallic waveguide cavity can be formed by two parallel metallic plates, both with a central elliptical-shaped aperture. A rigid acrylic plastic tube with an elliptical cross section is attached to the parallel plates and is used to hold water within the metallic waveguide cavity section, keeping the mass of the water from resting on the patient. In this conceptual example with eight monopole radiating elements, the phased array transmitter would distribute eight coherent signals and use eight adaptively controlled phase shifters and power amplifiers to deliver the desired phase values $(\phi_1, \phi_2, \cdots, \phi_8)$ and power values $(p_1, p_2, \cdots, p_8)$ to the monopole array elements as depicted in Figure 7.2. In this hyperthermia array applicator design, the monopole antenna elements are parallel to each other and are located at a fixed distance from the cylindrical metallic backwall of the parallel plate metallic waveguide cavity. The monopole elements are arranged in a ring and can be spaced typically about one-eighth to one-quarter of a wavelength from the reflecting ground plane behind each monopole element.

In Figure 7.2, the monopole array elements are each driven adaptively by variable RF phase shifters $(\phi_1, \phi_2, \cdots, \phi_8)$ and variable power amplifiers $(p_1, p_2, \cdots, p_8)$. RF signals such as continuous wave (CW) (oscillator), pulsed, ultrawideband, or other waveforms suitable for thermotherapy can be

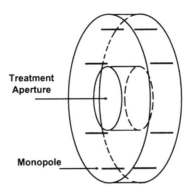

Figure 7.1 Conceptual monopole phased array thermotherapy applicator for deep heating.

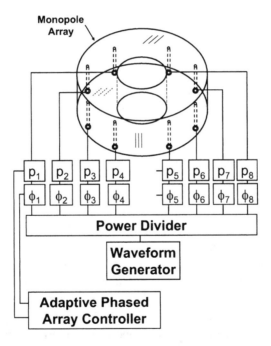

Figure 7.2 Block diagram for adaptive monopole phased array thermotherapy applicator with phase and power distribution network.

generated by a waveform generator, which divides into eight channels using a power divider.

A single metallic monopole array antenna element with length L and diameter D is depicted in Figure 7.3. The length L of a monopole antenna

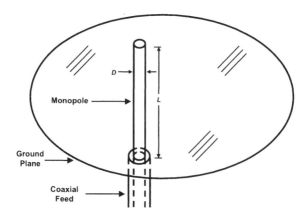

Figure 7.3 Diagram of a monopole array element with length L and diameter D.

element would typically be approximately one-quarter wavelength long. The monopole antenna element can be attached to the center conductor of a RF coaxial cable forming a coaxial feed aperture that illuminates the metallic conductor of the monopole antenna element.

The radiating monopole elements for the deep thermotherapy array are designed as follows. As discussed earlier, the wavelength in deionized water (used in the water bolus) is about 34 cm. In this design, the spacing between the monopoles and the cavity backwall is about 8 cm, and this spacing corresponds to approximately 0.235λ. The theoretical length of each monopole radiating antenna element is typically one-quarter wavelength, or approximately 8.5 cm. In the contruction of the monopole array, one can use microwave connectors and either solder a conducting rod to the center pin of the connector, or replace the center pin of the connector with a conducting (metallic) rod to form the monopole radiator. The diameter of the metallic rod monopole antenna element can be the same diameter as the center pin of a standard type-N coaxial connector, which is 0.3175 cm.

A schematic diagram of the monopole phased array applicator is shown in Figure 7.4. The elliptically shaped acrylic plastic tube has a thickness denoted t_p. A target body (defined by the parameters (a_2, b_2)) is positioned within the aperture of the monopole array applicator. A flexible water bolus, with inner and outer diameter parameters (a_2, b_2) and (a, b), respectively, is used to couple RF energy into the tissues of the patient. The monopole antenna element positions are located on a circle (ring) with radius R_a. The inner radius of the metallic waveguide housing is denoted R_w. The thickness of the metallic waveguide housing is denoted as t_w. The outer surface of the

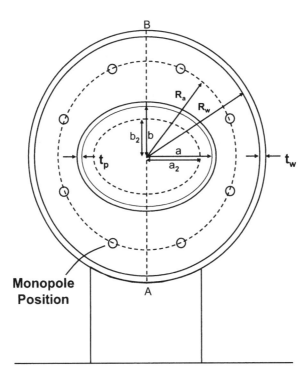

Figure 7.4 Diagram for a monopole phased array thermotherapy applicator with design parameters indicated.

metallic waveguide housing is supported by a rigid support member. Since the applicator is substantially rigid, only the mass of the water bolus applies pressure to the target body (patient). The cross-sectional opening of the patient treatment aperture can be approximately 42 to 52 cm wide by 30 to 38 cm high to accommodate most patients. The waveguide aperture opening (along the axial or longitudinal direction of the patient) is approximately one-half of a wavelength. At 100 MHz, the wavelength in water is approximately 34 cm, thus one-half of a wavelength is about 17 cm. The E-field radiation is confined to be no larger than this 17-cm longitudinal waveguide-aperture region. The aperture opening can actually vary from about one-third of a wavelength to over one-half of a wavelength. A flat or curved patient support can be made of a dielectric material, but is not shown here.

As discussed in Chapter 3, the specific absorption rate (SAR) is a parameter used in quantifying the heating performance of thermotherapy applicators. With proper choice of the ring array diameter, it is possible to reduce the level of surface SAR compared to the SAR produced at depth in the tumor or treatment region. Fundamentally, this effect is due to spherical wave (see (3.70)) versus plane wave (see (3.71)) radiation for the antenna radiating elements. A plane wave attenuates rapidly in muscle tissue due to the dielectric loss of the tissue. For spherical waves, in addition to dielectric losses the wave attenuates inversely proportional to the radial distance R. Plane waves penetrate deeper than spherical waves since the $1/R$ radial dependence of E-field attenuation with depth is not present. A plane wave is attenuated only by the loss due to dielectric material attenuation. A spherical wavefront from the radiating elements can be made to appear more planar (plane-wave like) at central depth by allowing the diameter of the ring array to grow. Thus, an effective larger diameter ring array may yield deeper penetration in tissue compared to a smaller diameter ring array. The monopole phased array waveguide design makes this larger ring array possible. Using the method of images as described by Jackson [55], in Figure 7.5, the reflecting surface behind the active radiating monopole elements provides an approximate secondary image array of monopoles with a resulting effective larger array diameter. Assuming the transmit monopoles are close to the waveguide cavity wall and assuming a flat conductor, the effective radius, denoted R_i, of the image monopole array is approximately equal to

$$R_i = 2R_w - R_a \tag{7.1}$$

A more accurate value for the image distance R_i, for a source antenna located within a cylindrical cavity, can be derived by enforcing boundary conditions as described by Jackson [55]. The approximate location of the effective phase center of the primary monopole array and the image monopole array is at the midpoint of the arrays or at range distance R_w from the origin.

In this example array design, the distance from the monopole to the reflecting backwall is 8 cm. Referring to Figure 7.5, assuming the radius of the monopole array is $R_a = 30$ cm and the reflecting wall surface has an approximate radius $R_w = 38$ cm, then the image array radius would be $R_i = 46$ cm. The image array has a diameter of 92 cm and would have less spherical wave relative attenuation in the tissue compared to the active array with diameter 60 cm.

Adaptive focusing for heating a deep tumor is a straightforward adjustment of the phase shift of each monopole element of the phased array so that the E-field is maximized forming a focus at the tumor. However, it is expected that adaptive focusing alone may not be adequate in general to avoid superficial hot spots. As demonstrated in Chapters 5 and 6, noninvasive adaptive nulling of the superficial fields can be accomplished using feedback from E-field sensors located on the skin surface at one or more null positions at the probes and by controlling the power and phase to each radiating monopole

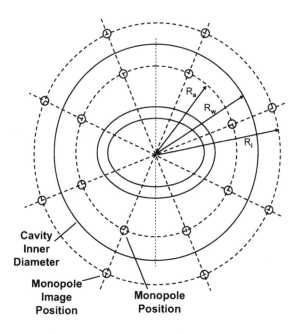

Figure 7.5 Diagram for a monopole phased array thermotherapy applicator with the method of images applied approximately in two dimensions to account for first-order reflection from the waveguide cavity wall. The inner ellipse represents a phantom target, and the surrounding ellipse represents the aperture of the monopole array.

antenna. The null zone surrounding each surface E-field sensor penetrates into the body and protects the skin and subcutaneous tissues. Demonstrations of adaptive nulling and deep heating in phantoms have been conducted, for example on a four-channel ring array of dipoles as discussed in Chapter 6.

The monopole phased array described above was analyzed using the finite-difference time-domain (FDTD) method according to the formulation described in Chapter 3. The monopole array heating performance is evaluated by calculating the SAR.

7.3 SIMULATION RESULTS

The finite-difference time-domain (FDTD) simulation results at 100 MHz for the monopole array design presented in the previous section are now considered. In this example, the monopole elements are located in a ring 60 cm in diameter as shown in Figure 7.6. The monopole array element coordinates are listed in Table 7.1. The monopole elements are surrounded

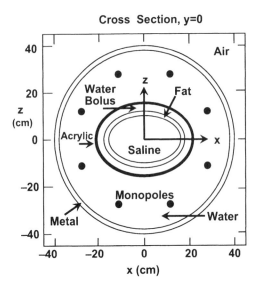

Figure 7.6 Cross-sectional view for an example monopole phased array thermotherapy applicator and elliptical phantom modeled using the three-dimensional finite-difference time-domain technique.

Table 7.1
Antenna Element Coordinates for Monopole Phased Array Thermotherapy Applicator Shown in Figure 7.6

Element Number	x (cm)	z (cm)
1	−11.5	27.7
2	11.5	27.7
3	27.7	11.5
4	27.7	−11.5
5	11.5	−27.7
6	−11.5	−27.7
7	−27.7	−11.5
8	−27.7	11.5

by a water-filled metallic cavity having the outer cylindrical-shaped surface of the cavity with an inner diameter of 76 cm and an outer diameter of 80 cm. The monopole elements are spaced 8 cm from the cavity backwall. The cylindrical-shaped cavity wall surrounding the radiating monopole elements is modeled as a highly conducting metal such as aluminum (dielectric constant 1.0, electrical conductivity 3.72×10^7 S/m). A side view of the monopole array

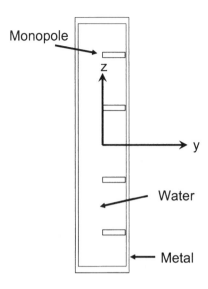

Figure 7.7 Projected midline-plane view for an example monopole phased array waveguide applicator.

projected onto the midline plane is shown in Figure 7.7 — due to symmetry only four monopole elements are shown in this projection. A side view of the thermotherapy applicator and elliptical phantom in the midline (median) plane (x=0) is shown in Figure 7.8. The medium surrounding the applicator and phantom is air (dielectric constant 1.0, conductivity 0.0 S/m)

   In the model of the water bolus, the dielectric constant of water is 78.0, and the electrical conductivity is 0.0001 S/m at 100 MHz [53, 54]. The phantom body (typical of muscle) is modeled by saline. The phantom salinity s in parts per thousand (ppt) (grams salt per kg water) is s= 9g/kg or 9 ppt, which is 0.9% NaCl in deionized water (dielectric constant 77.0, conductivity 0.5 S/m) [53, 54]. The outer 2 cm of the phantom is modeled by a uniform layer of fat (dielectric constant 7.0, conductivity 0.07 S/m at 100 MHz). The major axis of the elliptical phantom (including the fat layer) is 36 cm and the minor axis is 24 cm — this type of phantom has been used experimentally with an adaptive phased array applicator as described in Chapter 6.

   The finite-difference time-domain solution of the monopole phased array with phantom was modeled as a three-dimension volume consisting of 90 × 56 cubic cells in cross section by 90 cubic cells in the longitudinal dimension. Each cube is 1 cm on a side and the material dielectric properties were set equal to those of air, metal, water, muscle, fat, and acrylic as desired.

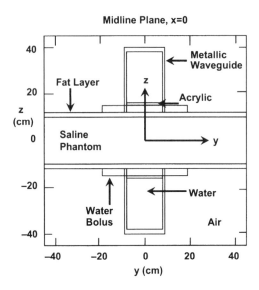

Figure 7.8 Midline-plane view for an example monopole phased array thermotherapy applicator and elliptical phantom modeled using the three-dimensional finite-difference time-domain technique.

The 3-cm space between the aperture of the monopole array applicator and the phantom is modeled by water (the water bolus). The applicator aperture is modeled with an acrylic plastic material (dielectric constant 2.55, conductivity 0.0008 S/m), which seals the aperture of the monopole array. The outer dimension of the major axis of the acrylic material filling the aperture is 44 cm and the minor axis is 32 cm. The outer dimension of the major axis of the water bolus is 42 cm and the minor axis is 30 cm. The outer dimension of the major axis of the phantom fat layer is 36 cm and the minor axis is 24 cm. The fat layer has a thickness of 2 cm. The major axis of the phantom saline region is 32 cm and the minor axis is 20 cm.

The longitudinal aperture dimension of the monopole array waveguide is 16 cm. The monopole radiators are modeled as seven cubic cells long, so they have a metallic length of 7 cm. The monopoles have a feed gap of one cubic cell or 1 cm. Thus, the distance from the distal tip of the monopole to the ground plane is 8 cm.

For calculation purposes, the excitation frequency was selected as 100 MHz, and the phase at each monopole was adjusted to focus the peak electric field at the midpoint of the phantom $(x = 0, y = 0, z = 0)$. The FDTD software was used to calculate the E-field amplitude and phase pattern

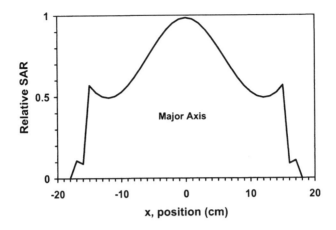

Figure 7.9 Finite-difference time-domain calculated SAR along the major axis of the elliptical saline phantom irradiated at 100 MHz by a monopole phased array thermotherapy applicator.

for each monopole radiating one at a time, and then a second computer program calculated (by superposition) the E-field radiation pattern and the specific absorption rate (SAR) pattern of the complete array. The graph shown in Figure 7.9 displays the calculated SAR along the major axis of the elliptical phantom at $y = 0, z = 0$. The graph in Figure 7.10 displays the calculated SAR along the minor axis of the elliptical phantom at $x = 0, y = 0$. Figure 7.11 is a graph displaying the SAR along the longitudinal axis of the phantom at $x = 0, z = 0$. The 50% SAR value is typically used as a measure of the effective heating zone for a hyperthermia applicator. The single peak along the major and minor axes indicates that the desired, simulated, adaptively focused deep-heating pattern is achieved. Further, the Gaussian-(bell) shaped SAR pattern along the longitudinal axis indicates that the 50% SAR is confined to about the width of the monopole array waveguide aperture (about 17 cm). A larger zone of heating in the longitudinal dimension is possible by providing two monopole array applicators separated by a distance s_y. The two longitudinal applicators can be fed coherently (with a common oscillator) or noncoherently (with separate oscillators).

7.4 SUMMARY

In this chapter, a radiofrequency monopole phased array applicator for deep focused heating has been described. The monopole phased array

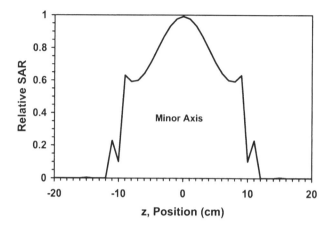

Figure 7.10 Finite-difference time-domain calculated SAR along the minor axis of the elliptical saline phantom irradiated at 100 MHz by a monopole phased array thermotherapy applicator.

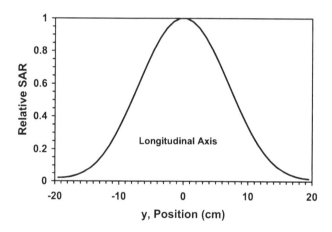

Figure 7.11 Finite-difference time-domain calculated SAR along the longitudinal axis of the elliptical saline phantom irradiated at 100 MHz by a monopole phased array thermotherapy applicator.

thermotherapy applicator RF performance was simulated using the three-dimensional finite-difference time-domain technique to compute the specific absorption rate in an elliptical saline-filled phantom. The simulated results demonstrate deep-heating characteristics for a saline-filled elliptical phantom. Further analysis and phantom testing of the monopole phased array applicator with adaptive array algorithms for deep tumor heating of the torso [56] and other areas of the body such as head and neck [57] is warranted.

References

[1] Lin, J.C., (ed.), "Special Issue on Phased Arrays for Hyperthermia Treatment of Cancer," *IEEE Trans. on Microwave Theory and Techniques,* Vol. MTT-34, No. 5, 1986.

[2] Sathiaseelan, V., M.F. Iskander, G.C.W. Howard, and N.M. Bleehen, "Theoretical Analysis and Clinical Demonstration of the Effect of Power Control Using the Annular Phased-Array Hyperthermia System," *IEEE Trans. on Microwave Theory and Techniques,* Vol. MTT-34, No. 5, 1986, pp. 514-519.

[3] Sathiaseelan, V., "Potential for Patient-Specific Optimization of Deep Heating Patterns Through Manipulation of Amplitude and Phase," *Strahlentherapie Onkologie,* Vol. 165, No. 10, 1989, pp. 743-745.

[4] Sullivan, D., "Three-Dimensional Computer Simulation in Deep Regional Hyperthermia Using the Finite-Difference Time-Domain Method," *IEEE Trans. Microwave Theory and Techniques,* Vol. MTT-38, No. 2, 1990, pp. 204-211.

[5] Sullivan, D., "Mathematical Methods for Treatment Planning in Deep Regional Hyperthermia," *IEEE Trans. Microwave Theory and Techniques,* Vol. 39, No. 5, 1991, pp. 864-872.

[6] Sato, G., C. Shibata, S. Sekimukai, H. Wakabayashi, K. Mitsuka, and K. Giga, "Phase-Controlled Circular Array Heating Equipment for Deep-Seated Tumors: Preliminary Experiments," *IEEE Trans. Microwave Theory and Techniques,* Vol. MTT-34, No. 5, 1986, pp. 520-525.

[7] Cudd, P.A., A.P. Anderson, M.S. Hawley, and J. Conway, "Phased-Array Design Considerations for Deep Hyperthermia Through Layered Tissue," *IEEE Trans. Microwave Theory and Techniques,* Vol. MTT-34, No. 5, 1986, pp. 526-531.

[8] Morita, N., T. Hamasaki, and N. Kumagai, "An Optimal Excitation Method in Multiapplicator Systems for Forming a Hot Zone Inside the Human Body," *IEEE Trans. Microwave Theory and Techniques,* Vol. MTT-34, No. 5, 1986, pp. 532-538.

[9] De Wagter, C., "Optimization of Simulated Two-Dimensional Temperature Distributions Induced by Multiple Electromagnetic Applicators," *IEEE Trans. Microwave Theory and Techniques,* Vol. MTT-34, No. 5, 1986, pp. 589-596.

[10] Knudsen, M., and U. Hartmann, "Optimal Temperature Control with Phased-Array Hyperthermia System," *IEEE Trans. Microwave Theory and Techniques,* Vol. MTT-34, No. 5, 1986, pp. 597-603.

[11] Boag, A., and Y. Leviatan, "Optimal Excitation of Multiapplicator Systems for Deep Regional Hyperthermia," *IEEE Trans. Biomedical Engineering,* Vol. BME-37, No. 10, 1990, pp. 987-995.

[12] Loane III, J.T., and S.W. Lee, "Gain Optimization of a Near-Field Focusing Array for Hyperthermia Applications," *IEEE Trans. Microwave Theory and Techniques,* Vol. 37, No. 10, 1989, pp. 1629-1635.

[13] Turner, P.F., T. Schaefermeyer, and T. Saxton, "Future Trends in Heating Technology of Deep-Seated Tumors," *Recent Results in Cancer Research*, Vol. 107, 1988, pp. 249-262.

[14] Turner, P.F., A. Tumeh, and T. Schaefermeyer, "BSD-2000 Approach for Deep Local and Regional Hyperthermia: Physics and Technology," *Strahlentherapie Onkologie*, Vol. 165, No. 10, 1989, pp. 738-741.

[15] Strohbehn, J.W., and R.B. Roemer, "A Survey of Computer Simulations of Hyperthermia Treatments," *IEEE Trans. on Biomedical Engineering,* Vol. BME-31, No. 1, 1984, pp. 136-149.

[16] Ocheltree, K.B., and L.A. Frizzell, "Determination of Power Deposition Patterns for Localized Hyperthermia: A Transient Analysis," *Int. J. Hyperthermia,* Vol. 4, No. 3, 1988, pp. 281-296.

[17] Shimm, D.S., T.C. Cetas, J.R. Oleson, E.R. Gross, D.N. Buechler, A.M. Fletcher, and S.E. Dean, "Regional Hyperthermia for Deep-Seated Malignancies Using the BSD Annular Array," *Int. J. Hyperthermia*, Vol. 4, No. 2, 1988, pp. 159-170.

[18] Myerson, R.J., L. Leybovich, B. Emami, P.W. Grigsby, W. Straube, and D. von Gerichten, "Phantom Studies and Preliminary Clinical Experience with the BSD 2000," *Int. J. Hyperthermia*, Vol. 7, No. 6, 1991, pp. 937-951.

[19] Wust, P., J. Nadobny, R. Felix, P. Deulhard, A. Louis, and W. John, "Strategies for Optimized Application of Annular-Phased-Array Systems in Clinical Hyperthermia," *Int. J. Hyperthermia*, Vol. 7, No. 1, 1991, pp. 157-173.

[20] Nikita, K.S., N. Maratos, and N.K. Uzunoglu, "Optimum Excitation of Phases and Amplitudes in a Phased Array Hyperthermia System," *Int. J. Hyperthermia*, Vol. 8, No. 4, 1992, pp. 515-528.

[21] Straube, W.L., E.G. Moros, and R.J. Myerson, "Phase Stability of a Clinical Phased Array System for Deep Regional Hyperthermia," *Int. J. Hyperthermia*, Vol. 11, No. 1, 1995, pp. 87-93.

[22] Paulsen, K.D., S. Geimer, J. Tang, and W.E. Boyse, "Optimization of Pelvic Heating Rate Distributions with Electromagnetic Phased Arrays," *Int. J. Hyperthermia*, Vol. 15, No. 3, 1999, pp. 157-186.

[23] Seebass, M., R. Beck, J. Gellermann, J. Nadobny, and P. Wust, "Electromagnetic Phased Arrays for Regional Hyperthermia: Optimal Frequency and Antenna Arrangement," *Int. J. Hyperthermia*, Vol. 17, No. 4, 2001, pp. 321-336.

[24] Wiersma, J., R.A.M. van Maarseveen, and J.D.P. van Dijk, "A Flexible Optimization Tool for Hyperthermia Treatments with RF Phased Array Systems," *Int. J. Hyperthermia*, Vol. 18, No. 2, 2002, pp. 73-85.

[25] Shi, G., and W.T. Joines, "Design and Analysis of Annular Antenna Arrays with Different Reflectors," *Int. J. Hyperthermia*, Vol. 20, No. 6, 2004, pp. 625-636.

[26] Kongsli, J., B.T. Hjertaker, and T. Frøystein "Evaluation of Power and Phase Accuracy of the BSD Dodek Amplifier for Regional Hyperthermia Using an External Vector Voltmeter Measurement System," *Int. J. Hyperthermia*, Vol. 22, No. 8, 2006, pp. 657-671.

[27] Jones, E., A.A. Secord, L.R. Prosnitz, T.V. Samulski, J.R. Oleson, A. Berchuck, D. Clarke-Pearson, J. Soper, M. W. Dewhirst, and Z. Vujaskovic, "Intra-Peritoneal Cisplatin and Whole Abdomen Hyperthermia for Relapsed Ovarian Carcinoma," *Int. J. Hyperthermia*, Vol. 22, No. 2, 2006, pp. 161-172.

[28] Fatehi, D., J. van der Zee, M. de Bruijne, M. Franckena, and G.C. van Rhoon, "RF-power and Temperature Data Analysis of 444 Patients with Primary Cervical Cancer: Deep Hyperthermia Using the Sigma-60 Applicator is Reproducible," *Int. J. Hyperthermia*, Vol. 23, No. 8, 2007, pp. 623-643.

[29] Fatehi, D., and G.C. van Rhoon, "SAR Characteristics of the Sigma-60-Ellipse Applicator," *Int. J. Hyperthermia*, published online January 9, 2008, (www.thermalmedicine.org).

[30] Mittal, B.B., et al., "Regional Hyperthermia in Patients with Advanced Malignant Tumors: Experience with the BSD 2000 Annular Phased-Array System and Sigma-60 Applicator," *Endocurietherapy/Hyperthermia Oncology*, Vol. 10, 1994, pp. 223-236.

[31] von Hippel, A.R., A.H. Runck, and W.B. Westphal, *Dielectric Analysis of Biomaterials*, Cambridge, MA: Laboratory for Insulation Research, Massachusetts Institute of Technology, Technical Report 13, AD-769 843, 1973.

[32] Fenn, A.J., "Adaptive Nulling Hyperthermia Array," US Patent No. 5,251,645, October 12, 1993.

[33] Fenn, A.J., "Adaptive Focusing and Nulling Hyperthermia Annular and Monopole Phased Array Applicators," US Patent No. 5,441,532, August 15, 1995.

[34] Fenn, A.J., "Non-Invasive Adaptive Nulling for Improved Hyperthermia Thermal Dose Distribution," *IEEE Engineering in Medicine and Biology Society Int Conf*, October 31 - November 3, 1991, Vol. 13, No. 2, 1991, pp. 976-977.

[35] Fenn, A.J., and G.A. King, "Adaptive Nulling in the Hyperthermia Treatment of Cancer," *The Lincoln Laboratory Journal*, Lincoln Laboratory, Massachusetts Institute of Technology, Vol. 5, No. 2, 1992, pp. 223-240.

[36] Fenn, A.J., C.J. Diederich, and P.R. Stauffer, "An Adaptive-Focusing Algorithm for a Microwave Planar Phased-Array Hyperthermia System," *The Lincoln Laboratory Journal*, Lincoln Laboratory, Massachusetts Institute of Technology, Vol. 6, No. 2, 1993, pp. 269-288.

[37] Fenn, A.J., and G.A. King, "Experimental Investigation of an Adaptive Feedback Algorithm for Hot Spot Reduction in Radio-Frequency Phased-Array Hyperthermia," *IEEE Trans Biomed Eng.*, Vol. 43, No. 3, 1994, pp. 273-280.

[38] Fenn, A.J., and G.A. King, "Adaptive Radio Frequency Hyperthermia Phased Array System for Improved Cancer Therapy: Phantom Target Measurements," *Int J Hyperthermia*, Vol. 10, No. 2, 1994, pp. 189-208.

[39] Fenn, A.J., B.A. Bornstein, G.K. Svensson, and H.F. Bowman, "Minimally Invasive Monopole Phased Arrays for Hyperthermia Treatment of Breast Carcinomas: Design and Phantom Tests," *Int. Symp. on Electromagnetic Compatibility*, Sendai, Japan: Vol. 10, No. 2, 1994, pp. 566-569.

[40] Fenn, A.J., "Minimally Invasive Monopole Phased Arrays for Hyperthermia Treatment of Breast Cancer," In: *Proc. 1994 Int. Symp. on Antennas,* Nice, France: November 8-10, 1994, pp. 418-421.

[41] Fenn, A.J., "Minimally Invasive Monopole Phased Array Hyperthermia Applicators and Method for Treating Breast Carcinomas," US Patent No. 5,540,737, July 30, 1996.

[42] Fenn, A.J., V. Sathiaseelan, G.A. King, and P.R. Stauffer, "Improved Localization of Energy Deposition in Adaptive Phased-Array Hyperthermia Treatment of Cancer," *The Lincoln Laboratory Journal,* Lincoln Laboratory, Massachusetts Institute of Technology, Vol. 9, No. 2, 1996, pp. 187-196.

[43] Sathiaseelan, V., A.J. Fenn, and A. Taflove, "Recent Advances in External Electromagnetic Hyperthermia," In: Chapter 10 of *Advances in Radiation Treatment,* Mittal, B.B., J.A. Purdy, and K.K. Ang, (eds.), Boston, Massachusetts: Kluwer Academic Publishers, 1998, pp. 213-245.

[44] Fenn, A.J., V. Sathiaseelan, G.A. King, and P.R. Stauffer, "Improved Localization of Energy Deposition in Adaptive Phased Array Hyperthermia Treatment of Cancer," *J Oncol Management,* Vol. 7, No. 2, 1998, pp. 22-29.

[45] Fenn, A.J., "Thermodynamic Adaptive Phased Array System for Activating Thermosensitive Liposomes in Targeted Drug Delivery," US Patent No. 5,810,888, September 22, 1998.

[46] Fenn, A.J., "Monopole Phased Array Thermotherapy Applicator for Deep Tumor Therapy," US Patent No. 6,807,446, October 19, 2004.

[47] Fenn, A.J., G.L. Wolf, and R.M. Fogle, "An Adaptive Phased Array for Targeted Heating of Deep Tumors in Intact Breast: Animal Study Results," *Int J Hyperthermia,* Vol. 15, No. 1, 1999, pp. 45-61.

[48] Gavrilov, L.R., J.W. Hand, J.W. Hopewell, and A.J. Fenn, "Pre-clinical Evaluation of a Two-Channel Microwave Hyperthermia System with Adaptive Phase Control in a Large Animal," *Int J Hyperthermia,* Vol. 15, No. 6, 1999, pp. 495-507.

[49] Gardner, R.A., H.I. Vargas, J.B. Block, C.L. Vogel, A.J. Fenn, G.V. Kuehl, and M. Doval, "Focused Microwave Phased Array Thermotherapy for Primary Breast Cancer," *Ann Surg Oncol,* Vol. 9, No. 4, 2002, pp. 326-332.

[50] Vargas, H.I., W.C. Dooley, R.A. Gardner, K.D. Gonzalez, S.H. Heywang-Kobrunner, and A.J. Fenn, "Focused Microwave Phased Array Thermotherapy for Ablation of Early-Stage Breast Cancer: Results of Thermal Dose Escalation," *Ann Surg Oncol,* Vol. 11, No. 2, 2004, pp. 139-146.

[51] Fenn, A.J., *Breast Cancer Treatment by Focused Microwave Thermotherapy,* Sudbury, MA: Jones and Bartlett, 2007.

[52] Vargas, H.I., W.C. Dooley, A.J. Fenn, M.B. Tomaselli, and J.K. Harness, "Study of Preoperative Focused Microwave Phased Array Thermotherapy in Combination With Neoadjuvant Anthracycline-Based Chemotherapy for Large Breast Carcinomas," *Cancer Therapy,* Vol. 5, 2007, pp. 401-408, published online (www.cancer-therapy.org), November 25, 2007.

[53] Stogryn, A., "Equations for Calculating the Dielectric Constant of Saline Water," *IEEE Trans Microwave Theory and Techniques*, Vol. 19, No. 8, 1971, pp. 733-736.

[54] Malmberg, C.G., and A.A. Maryott, "Dielectric Constant of Water from 0 to 100°C," *J Res National Bureau of Standards,* Vol. 56, No. 1, 1956, pp. 1-8.

[55] Jackson, J.D., *Classical Electrodynamics*, (2nd ed.), New York: John Wiley, 1975, pp. 54-58.

[56] Fenn, A.J., "Adaptive Thermodynamic Therapy (TDT) Drug Delivery System for Treating Deep-Seated Cancer," *Drug Delivery Technology*, October 2002, pp. 74-79.

[57] Fenn, A.J., D.S. Poe, C.E. Reuter, A. Taflove, "Noninvasive Monopole Phased Array for Hyperthermia Treatment of Cranial-Cavity and Skull-Base Tumors: Design, Analysis, and Phantom Tests," *Proc of the 15th Annual IEEE Engineering in Medicine and Biology Society International Conference,* October 28-31, 1993, Part 3, pp. 1453-1454.

8

Adaptive Array for Breast Cancer: Preclinical Results

8.1 INTRODUCTION

An estimated 182,460 new cases of invasive breast cancer are expected to be diagnosed in 2008 in the United States [1]. Breast cancer typically begins in the ductal system (milk ducts and glandular lobules) and becomes invasive once the cancer cells penetrate the ductal or glandular wall [2]. Current treatments for breast cancer include surgery, radiation therapy, chemotherapy, and hormonal therapy depending on the stage of the cancer as described by Newman [3] and Carlson [4]. For early-stage breast cancer, the tumor is surgically excised prior to radiation therapy. Chemotherapy is sometimes administered prior to surgery to shrink large breast cancer tumors. Although these treatments generally are effective, in many cases after the treatments are completed, residual cancer cells remain in the breast, which can lead to a recurrence of the cancer as described by Singletary [5], and there are numerous treatment side effects. Breast surgery can involve either breast conservation, which can be described as a wide-excision lumpectomy (also referred to as a partial mastectomy or quadrantectomy), or it can involve a mastectomy in which the entire breast is removed. Overall survival at 20 years post treatment is generally equivalent among patients receiving total mastectomy (47%), lumpectomy (46%), or lumpectomy plus radiation therapy (47%) as described by Fisher [6]. Mastectomy rates have risen in the last 3 years based on a study published by Katipamula [7] — at the Mayo Clinic in Rochester, MN, the rate of patients receiving mastectomy was 43% in 2006 compared to 30% in 2003. The increase in the rate of mastectomy was correlated with an increased use of magnetic resonance imaging (MRI) of the breast, 22% in 2006 compared to 11% in 2003. *Treatments that could reduce the rate of mastectomy, increase the use of breast conservation, and eliminate residual cancer cells are desirable.*

This chapter describes characteristics of a wide-treatment-field adaptive phased array focused microwave thermotherapy system for treating breast cancer tumors in the intact breast [8-32]. A detailed discussion of potential protocols for using focused microwave thermotherapy for treating breast cancer has been given by the author [29, pp. 183-202]. Surgical considerations for breast cancer are discussed in Section 8.2. Microwave considerations for breast tissue and breast cancer are discussed in Section 8.3. A detailed description of the adaptive phased array thermotherapy system concept for breast cancer is given in Section 8.4. In Section 8.5, the mathematical formulation for an adaptive phased array algorithm for a focused microwave thermotherapy system is given. Ray-tracing calculations for the specific absorption rate (SAR) delivered to breast tissue by a 915-MHz focused microwave phased array are given in Section 8.6. After describing the basic design in Sections 8.2 to 8.6, in Section 8.7 a clinical adaptive phased array focused microwave thermotherapy system for treating breast cancer in the intact breast is described. In Section 8.8, the finite-difference time-domain technique is used to compute the specific absorption rate delivered by a focused microwave waveguide phased array for simulated breast tissue and simulated breast tissue with tumors. The measured SAR for a compressed breast phantom with variable size tumors heated by an adaptive focused microwave phased array thermotherapy system is given in Section 8.9. Animal trial testing with the adaptive microwave phased array thermotherapy system is briefly described in Section 8.10. Section 8.11 contains the summary.

8.2 SURGICAL CONSIDERATIONS FOR BREAST CANCER

The structure of the breast is described in detail by Haagensen [33]. The breast consists of epithelial parenchyma (cellular layer) of acini (milk-producing cells) and ducts and their supporting muscular and fascial elements (connective tissue), adipose tissue (fat), blood vessels, nerves, and lymphatics. The epithelial parenchyma consists of 20 or more lobes, which are each divided into numerous lobules (gland fields) each made up of 10 to 100 or more acini grouped around a collecting duct. The acini in the resting mammary gland are lined by a single layer of epithelial cells. An occasional second layer of epithelial cells around the base of the acinus is referred to as a myoepithelial cell layer. Myoepithelial cells of the breast resemble smooth muscle cells and take part in certain benign proliferations. The myoepithelial cells provide a muscular mechanism for ejecting milk from the acini and ducts. The bulk of the subareolar area and nipple is made of smooth muscle fibers – when the muscle fibers in the nipple contract, they empty the milk sinuses.

In many cases, breast cancer can be considered to be composed of a primary invasive cancerous tumor and a diffuse component of microscopic cancer cells that invade surrounding breast tissues. Breast conservation surgery attempts to excise all of the cancerous tumor cells. During the breast-conserving surgical procedure, for example during a first incision, the excised breast tissue is sent for intraoperative pathologic evaluation. The pathology procedures determine if there are tumor cells at the surgical cut edge, which is referred to as a positive tumor margin. Multicolored inking is used to identify specific margins as described by Newman [3] and Singletary [5]. If the tumor margin (or margins) are positive, the surgeon often will excise additional breast tissue in the region where the positive margin occurred, and this additional surgery is referred to as a reexcision. If tumor cells are found close (typically within 1 mm) to the surgical cut edge of the excised specimen, the surgeon will likely perform a reexcision to provide negative margins (negative margins are defined here as tumor cells greater than 1 mm from the surgical cut edge). There is a significantly higher cancer recurrence rate when positive tumor margins occur compared to when the margins are negative. For example, when the negative margin width is taken as >1 mm, five studies reviewed by Singletary [5, Table 1] of 1375 patients with invasive breast cancer receiving breast conservation therapy, with 57 to 127 months follow-up, showed that the mean locoregional recurrence rate for positive margins was 16.2% versus 2.6% for negative margins.

In the case of a wide-excision lumpectomy, the primary tumor and a 2- to 3-cm margin of tissue surrounding the tumor are excised. Therefore, even for a small early-stage tumor that is, say, 1 cm diameter, taking account of the 2- to 3-cm margin of tissue excised, the excised breast tissue volume can range from 5 to 7 cm in the maximum dimension. Similarly, a 3-cm tumor could require an excision with a 7- to 9-cm maximum diameter to achieve negative margins. Postoperative radiation therapy can be applied by either external or interstitial means as described by Arthur [34], and it attempts to eliminate any residual cancer cells that might be missed by the surgical procedure. For large primary breast cancer tumors, say 3 to 4 cm or larger, in clinical diameter, preoperative chemotherapy is sometimes used to provide local tumor shrinkage as well as to systemically treat cancer metastases (spread of cancer to other organs) as described by Hamilton [35], Kaufmann [36], Kaufmann [37], Fisher [38, 39], and Wolmark [40]. When preoperative chemotherapy (also referred to as neoadjuvant chemotherapy) successfully shrinks the large tumor, breast conservation surgery can be considered. An additional treatment modality that could provide increased tumor cell kill (ablation) and increased tumor shrinkage would need to treat large volumes of breast tissue in order to address small and large breast tumors and the tumor margins. A number of technologies including interstitial radiofrequency, interstitial laser, external focused ultrasound, interstitial cryotherapy, and wide-treatment-field adaptive phased array focused microwave thermotherapy are being developed and evaluated in clinical studies for ablation of breast cancer.

8.3 MICROWAVE THERMOTHERAPY CONSIDERATIONS FOR BREAST CANCER

The subject of breast cancer tumor thermal ablation as part of a multimodality approach in the treatment of breast cancer has been reviewed by Hall-Craggs [41], Singletary [42], Noguchi [43], Agnese and Burak [44], Huston and Simmons [45], and van Esser [46]. Thermal energy treatment with radiofrequency [47-52], interstitial laser photocoagulation [53, 54], focused ultrasound [55-58], cryotherapy [59, 60], or adaptive phased array focused microwave thermotherapy [26-31] has demonstrated some success in achieving ablation of breast cancer tumors. For breast cancer thermotherapy, microwave energy is promising because it preferentially heats and damages high-water high-ion content breast carcinomas, compared to lesser degrees of heating that occurs in lower-water, lower-ion content normal fatty breast tissues [61-69] as described below.

Based on a study by Campbell and Land [64], the percent water content of normal and malignant breast tissues has been measured. For example, the mean water content of normal fatty breast tissue is 19.1% ($n = 36$). Mixed glandular and connective tissue have a measured mean water content of 58.5% ($n = 20$). Breast carcinomas have a mean water content of 74.3% ($n = 22$). Note that these percent water content values are consistent with the data presented in Table 3.1. Based on these measured data for percent water content, the author makes the assumption that microwave energy might heat breast carcinomas faster than normal fatty breast tissue and glandular and connective tissues. As discussed below, a majority of the breast tissue are low-water content fatty breast tissue, with a lessor precentage of glandular and connective tissues.

To quantify the amount of normal fatty breast tissue in the breast, consider the following data. During 2000 to 2004, the median age for which breast cancer was diagnosed was 61 years [2]. Therefore, 50% of women that developed breast cancer were 61 years or younger and 50% were older than 61 years. The mean fraction by mass for the glandular content of the female breast decreases with patient age, and correspondingly the breast tissue contains more adipose (fatty) tissue as investigated by Klein [70]. The study by Klein showed that the mean glandular content drops linearly from about 70% to 40% as the patient age increases from 20 years to about 40 years. From age 40 years, the mean fraction by mass of glandular tissue reduces from 40% to about 35% by age 50, and then reduces to about 30% from age 55 where it remains essentially constant. Therefore, for women age 55 years and older, the average amount of normal fatty breast tissue is on the order of 70%. For purposes of analyzing the focused microwave heating of the breast and breast tumor at 915 MHz, the author assumes that for a majority of breast cancer patients, that the breast can be modeled as primarily normal fatty tissue. The dielectric properties of breast tissue are now considered.

Microwave radiation in the Industrial, Scientific, Medical (ISM) allocated frequency band, from 902 to 928 MHz, is commonly used in clinical hyperthermia systems and is the primary frequency band considered here. Microwave dielectric parameter data of female breast tissues from a few published studies exist at 915 MHz, which is the desired operating frequency for the focused microwave breast thermotherapy treatment described in this chapter; measured data suggest that carcinomas of the breast can be selectively heated compared to surrounding normal fatty breast tissues. The heatability of tissue is directly related to the ionic conductivity in the equation expressing the specific absorption rate given by (3.69).

Several articles with measured microwave-parameter (dielectric constant and electrical conductivity) data for breast tissues have been published by a number of authors, primarily Chaudhary [61] in 1984, Joines [62] in 1994, Surowiec [63] in 1988, Campbell and Land [64] in 1992, Burdette [65] in 1982, and more recently Lazebnik [66, 67] in 2007. A review article by Sha [68] in 2002 comparing the results of studies of the dielectric properties of normal and malignant breast tissue has also been published [68]. The article by Burdette provides measured microwave data for breast tissue at 500, 918, and 2450 MHz ($n = 3$ measurements per frequency); however, these data represent a small sampling and were measured through the skin and might not be completely representative of breast tissue itself (because of the possible effects of the intervening skin on the measurements) – the 918-MHz measured dielectric constant was 38.0 and the ionic conductivity was 0.6 S/m. The Campbell and Land article has measured water content data, and has measured microwave dielectric properties data only for 3.2-GHz microwaves. The dielectric properties of a wide variety of human tissues are described by Gabriel [69].

As discussed in Chapter 3, tissue dielectric properties are usually quantified in terms of dielectric constant and electrical (ionic) conductivity [69], and they are quantified for normal breast tissue and breast cancer from studies by Chaudhary [61] and Joines [62] in Table 8.1 and Table 8.2 for 915 MHz microwaves. The median values of dielectric constant and ionic conductivity for different percentages of adipose breast tissue are summarized in Table 8.3 for 915 MHz microwaves, based on large-scale measurements of breast surgery specimens by Lazebnik [67]. The measured data of Lazebnik show that normal breast tissue with low amounts (0% to 30%) of adipose tissue, have larger dielectric constant and conductivity $\epsilon_r = 48.9$ and $\sigma = 0.86$, which corresponds primarily to glandular and fibroconnective tissue. As mentioned earlier in Section 8.2, a muscular structure is used in ejecting milk from the acini and ducts, and this muscle component would likely be the source for the larger values of dielectric constant and ionic conductivity in the glandular tissue. From Table 3.3, muscle has a measured dielectric constant $\epsilon_r = 55$ and ionic conductivity $\sigma = 0.94$, which is very similar to the value for glandular and fibroconnective tissue measured by Lazebnik [67]. Comparing the results in Tables 8.1 and 8.3 it is clear that the measured data for Chaudhary and Joines are not for purely adipose breast tissue and the term that is used here designated *normal fatty breast tissue,* should be used in the sense of *containing a dominant amount of adipose tissue compared to other tissue such as glandular tissue in the breast.*

The relative dielectric constant primarily affects the wavelength of the microwaves propagating through tissue, and the electrical conductivity primarily affects the attenuation and power deposition of the microwaves

Table 8.1
Measured Dielectric Constant and Ionic Conductivity for Normal Fatty Breast Tissue at 915 MHz

Normal Fatty Breast Tissue, Measured	Dielectric Constant ϵ_r	Ionic Conductivity $\sigma(S/m)$
Chaudhary [61]	10	0.24
Joines [62]	15	0.18
Average	12.5	0.21

Table 8.2
Measured Dielectric Constant and Ionic Conductivity for Breast Cancer at 915 MHz

Breast Cancer Measured	Dielectric Constant ϵ_r	Ionic Conductivity $\sigma(S/m)$
Chaudhary [61]	60	0.93
Joines [62]	57.1	1.16
Average	58.6	1.03

in tissue. Increasing the dielectric constant of the tissue reduces the wavelength, and increasing the electrical conductivity increases the attenuation of microwaves. At 915 MHz, averaging the data from the Chaudhary (n = 15 measurements, 25°C) and Joines (n = 12 measurements, 24°C) articles, the average dielectric constant of normal fatty breast tissue is 12.5, and the average ionic conductivity is 0.21 S/m. In contrast, for breast cancer tumors the average dielectric constant is 58.6 and the average ionic conductivity is 1.03 S/m. Note that taking the ratio of the ionic conductivities for breast cancer tumor (Table 8.2) and normal breast tissue with 0% to 30% adipose tissue (Table 8.3), or $1.03/0.86 = 1.2$, suggests that there is about a 20% faster heating rate for breast cancer tumors compared to the heating of normal breast tissue with low amounts of adipose tissue. The dielectric parameter values for normal fatty breast tissue and breast cancer tumors are used in the ray-tracing and finite-difference time-domain simulations presented in this chapter. Note that the data from the Chaudhary and Joines studies are measured at room temperature (24° or 25°C). Also note that as the temperature of tissue increases, generally the dielectric constant decreases and the electrical conductivity increases. In summary, the dielectric parameters of normal fatty breast tissues and breast cancer tumors are similar to low-water-content fatty tissue and high-water-content muscle tissue, respectively.

8.4 ADAPTIVE PHASED ARRAY THERMOTHERAPY SYSTEM CONCEPT FOR BREAST CANCER

To heat deep tissues reliably at radiofrequency or microwave frequencies, it is necessary to surround the tissue with two or more coherent applicators controlled by an adaptive phased array algorithm [8-32]. An application to treat tumors in intact breast with an adaptive microwave phased array thermotherapy system is depicted in Figure 8.1 [24]. A two-channel adaptive phased array is used to heat deep tissues within a compressed breast [18, 19]

Table 8.3
Median Dielectric Constant and Ionic Conductivity Based on Measurements at 915 MHz for Various Percentages of Adipose Breast Tissue [67]

Percent Adipose Tissue (Range)	Median Dielectric Constant ϵ_r	Median Ionic Conductivity $\sigma(S/m)$
85 to 100%	4.82	0.044
31 to 84%	39.6	0.642
0 to 30%	48.9	0.86

Source: Lazebnik [67, Equation 1 p. 2644 and Table 3 p. 2650].

similar to the geometry used in X-ray mammography. Breast compression has a number of potential advantages for intact breast thermotherapy treatments. Utilization of breast compression results in less penetration depth required to achieve microwave heating at central depth. Compressing the breast to a flat surface improves the interface and electric-field coupling between the microwave applicator and the breast tissue, and allows a single pair of applicators to treat a wide range of breast sizes. Cooling of the breast compression plates and skin during thermotherapy treatments avoids the potential for skin-surface hot spots. Compressing the breast with the patient in a prone position, such as that used in 20- to 40-minute stereotactic needle

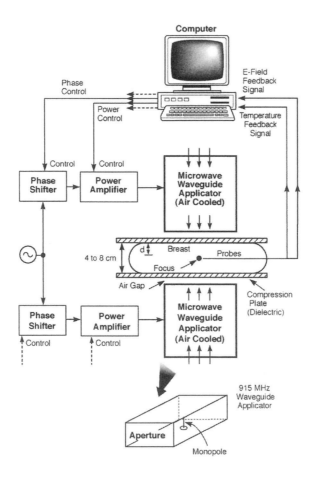

Figure 8.1 Block diagram for an adaptive phased array antenna focused microwave thermotherapy system for treating breast cancer. (From [24] with permission from Informa Healthcare, www.informaworld.com.)

breast biopsy procedures as discussed by Bassett [71] maximizes the amount of breast tissue within the compression device. Mild compression immobilizes the breast tissue such that any potential patient motion complications are eliminated. The compression device, which can include an aperture, is compatible with X-ray and ultrasound imaging techniques to accurately locate the tumor and assist in the placement of invasive E-field and temperature probe sensors. The amount of compression can be varied typically from about 4 to 8 cm to accommodate patient tolerance during a 20 to 40 minute or longer thermotherapy session. A patient-comfort study during breast compression in mammography shows that the mean compression thickness is 4.63 cm with a standard deviation (1 sigma) of 1.28 cm as described by Sullivan [72]. The study showed that mammography was painful (defined as either very uncomfortable or intolerable) in only 8% of the 560 women examined. Thus, hyperthermia treatments under mild breast compression for 20 to 40 minutes or longer should be feasible. Figure 8.2 shows a conceptual projected view of the breast compression plate, compressed breast, E-field and temperature probe locations, and waveguide applicator for breast thermotherapy treatments.

Prior to the hyperthermia treatment, a closed-end plastic catheter is inserted under ultrasound guidance within the breast tumor. In Figure 8.2, the window in the rigid dielectric (electromagnetically transparent material such as acrylic) compression plate is provided to allow an ultrasound transducer to contact the breast and guide the insertion of the catheter. Prior to inserting the catheter, a local anesthetic is applied to the skin entry point and the skin

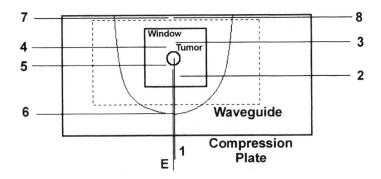

Figure 8.2 Diagram showing a projected view of breast compression plate, compressed breast, probe locations, and waveguide applicator for an adaptive phased array antenna focused microwave thermotherapy system for treating breast cancer.

is nicked. A combination E-field feedback sensor (Probe E in Figure 8.2) and fiber-optic temperature sensor (Probe 1 in Figure 8.2) are inserted within the catheter in the breast tumor. In Figure 8.2, additional temperature probes labeled 2 through 8 can be used to monitor surface temperatures of the breast. The E-field probe is used in monitoring the focal E-field amplitude as the phase shifters shown in Figure 8.1 are adjusted for maximum feedback signal using an adaptive phased array gradient-search algorithm as discussed in Section 8.5. The temperature probe within the breast tumor monitors the tumor temperature, which provides a feedback signal to control the output level of the power amplifiers during treatments. With a single 915-MHz microwave applicator, the depth of heating is unfocused and there is a tendency to overheat superficial tissues in order to heat deeper regions. In contrast, with a dual-applicator adaptive phased array, the E-field feedback probe allows the phase shifters to be adjusted so that a concentrated E-field can be generated, permitting focused heating in tissue at depth. In clinical treatments, microwave power amplifiers, phase shifters, cables, connectors, and associated control electronics of a microwave phased array hyperthermia system can have variations with time and temperature that can create large phase errors. Furthermore, body tissues are not homogeneous and they have electrical properties that can vary significantly with time and temperature. Thus, to take full advantage of coherent phased array electromagnetic radiating sources as described in the next section, it is essential to control adaptively the relative phase shift between the radiating sources based on the signal amplitude received at an E-field feedback probe located at the tumor site.

8.5 ADAPTIVE PHASED ARRAY ALGORITHM FOR FOCUSED MICROWAVE THERMOTHERAPY

A general-purpose multichannel adaptive phased array algorithm based on a gradient-search technique for controlling microwave power and phase during thermotherapy treatments has been developed [8, 14]. Based on the gradient-search algorithm derived in Section 2.3, an adaptive phased array focusing algorithm for rapidly adjusting the relative phase between two channels is described in this section. In a two-channel focused microwave breast thermotherapy system (Figure 8.1), only one phase shifter requires a relative phase (timing) adjustment, with respect to the other phase shifter, to achieve a microwave-focused condition.

As the relative microwave phase, denoted ϕ, changes over $\pm 180°$, the specific absorption rate in the tumor varies from a minimum to a maximum, with the minimum corresponding to a completed unfocused condition and the maximum specific absorption rate (SAR) corresponding to a focused condition. The specific absorption rate is proportional to the electric field squared (refer to (3.69)) or, equivalently, the microwave power measured by an electric-field probe positioned within a tumor. Let the measured microwave power received by the microwave probe be denoted as p^{rec}. A slow, brute-force, focusing approach would be to adjust the relative microwave phase in small phase increments (for example, 3 degrees) over the full 360 degrees, and then measure the received microwave power p^{rec} at each phase increment. For example, for an arbitrary breast tissue thickness and arbitrary tumor depth, if the phase increment (denoted $\Delta\phi$) is 3 degrees, it would take up to 120 phase adjustments and 120 power measurements to decide when the SAR maximum occurs. When the measured microwave power p^{rec} is a maximum, the corresponding phase at which the maximum occurs is the desired phase-focusing condition. Such a brute-force focusing approach can be slow because of the time required to adjust the phase and then measure the corresponding received microwave power.

A faster approach for focusing is a method of steepest ascent gradient-search algorithm that automatically searches for the correct phase setting that maximizes the received microwave power. Because of electromagnetic interference and noise, the measurement of microwave power by the probe in the tumor for small phase increments can have random fluctuations that could cause the algorithm to converge slowly and take at least 50 to 100 iterations to converge.

To reduce random fluctuations, it is possible to perform a series of microwave power measurements and average the data to improve the accuracy of the power measurement with a corresponding longer time to perform the phase-focusing operation and a reduction in the number of required iterations. Another method to speed the convergence of the gradient search is to include a fast acceleration term as described in Chapter 2, which can reduce the number of iterations to just a few (less than about 5 primary iterations). A fast acceleration gradient search for a two-channel adaptive phased array system is described mathematically as follows. With two channels, only the relative phase between the channels is important, so effectively only one adaptive phase channel is needed.

Let there be a series of primary iterations (microwave phase settings) with index ranging from 1 to J, and let j be the index at the jth primary iteration. Furthermore, for each primary iteration, let there be a subiteration index denoted k. At the jth primary iteration, the relative change (gradient) in received microwave power Δp_j^{rec} is computed, in terms of a finite difference, as the phase shifter setting ϕ is dithered up and down in phase ($\pm\Delta\phi$) and the absolute power is measured for each case, or

$$\Delta p_j^{rec} = p_j^{rec}(\phi_j + \Delta\phi) - p_j^{rec}(\phi_j - \Delta\phi) \tag{8.1}$$

If $\Delta p_j^{rec} > 0$, then to maximize the received microwave power the microwave phase should be adjusted in the positive phase direction. If $\Delta p_j^{rec} < 0$, then to maximize the received microwave power the microwave phase should be adjusted in the negative phase direction. Thus, the required gradient-search direction, denoted r, can be computed from (2.54) as

$$r = \Delta p_j^{rec}/|\Delta p_j^{rec}| \tag{8.2}$$

where the symbols $| \cdot |$ represents absolute value.

Based on (8.2), r takes on the value of either plus one $(+1)$ or minus one (-1) to indicate the desired search direction for the microwave phase to focus the array. Thus, for the adaptive phased array algorithm during the jth iteration of the phase adjustment, once the desired search direction (± 1 in this instance) is determined, the new phase increment during the subiterations ($k = 1, 2, 3, \cdots$) can take on the series of values $\Delta\phi, 2\Delta\phi, 4\Delta\phi, \cdots, 2^{k-1}\Delta\phi$. During these subiterations, the phase is commanded to take on the values

$$\phi_{j,k} = \phi_j + r2^{k-1}\Delta\phi, k = 1, 2, 3, \cdots \tag{8.3}$$

and during these subiterations while the new phase is set the microwave power received at the microwave probe in the tumor is measured and compared to the previous value. If $p_{j,k+1}^{rec} > p_{j,k}^{rec}$, then the microwave received power is increasing and the focus is improving, so the subiteration procedure continues. If $p_{j,k+1}^{rec} < p_{j,k}^{rec}$, then the microwave received power is decreasing and the focus is not improving, so the subiteration procedure stops and the main iteration index j is then incremented to $j + 1$ and the algorithm continues resuming with (8.1) to (8.3) until the maximum value of received microwave power is determined. A typical value for the phase increment $\Delta\phi$ can be, say, 5 degrees throughout the entire focusing algorithm, or this value can be changed (decreased) depending on the overall system design, taking account of the quantization and random errors in measuring the received microwave power as well as in setting the desired phase shifter values.

8.6 CALCULATED SPECIFIC ABSORPTION RATE FOR BREAST PHANTOM

As derived in Chapter 3 (see (3.69)), the microwave energy absorption per unit mass (or specific absorption rate, SAR, with units of W/kg) of tissue is directly proportional to both the electrical conductivity σ of the tissue and the square of the magnitude of the applied electric field, denoted |E|, and is inversely proportional to the tissue density (denoted ρ with units of kg/m^3):

$$SAR = \frac{1}{2}\frac{\sigma}{\rho}|E|^2 \tag{8.4}$$

As derived in Chapter 4 (see (4.30)), the rise in tissue temperature (ΔT) in a given time interval (Δt) with a given applied SAR is expressed as:

$$\Delta T = \frac{1}{c}SAR\Delta t \tag{8.5}$$

where c is the specific heat capacity of the tissue, and thermal conduction and perfusion effects are ignored. The SAR and rise in temperature defined by (8.4) and (8.5), respectively, are usually considered in terms of macroscopic heating of a solid tumor mass of a given size. However, one can also consider the above two equations on a microscopic level as being valid for the case of heating a single microscopic tumor cell. For now, a homogeneous phantom composed of simulated normal fatty breast tissue is considered as below.

Assume, for example, that a 6-cm homogeneous breast tissue phantom has the dielectric parameters as specified in Table 3.3 for normal fatty breast tissue. In this case, with dielectric constant $\epsilon_r = 12.5$ and electrical conductivity $\sigma = 0.21$ S/m, the attenuation of a 915-MHz plane wave is about 1 dB per centimeter and the propagation phase constant is 0.69 radians/cm. Using this value for the phase constant, the wavelength in normal fatty breast tissue is calculated from (3.36) to be 9.1 cm. Assume that there are two opposing 915-MHz plane waves (left and right) irradiating the breast phantom as in Figure 8.3. Let the focal point be defined as the central depth of the phantom; that is, the desired focal point is 3 cm beneath the skin surfaces (points A and B). Thus, to reach the central focal point, each plane wave must travel through 3 cm of tissue. The plane-wave electric fields due to applicators 1 and 2 are given by ray tracing in phasor form as

$$E_1(x) = E_o e^{-\alpha d_1} e^{-j\beta x} e^{j\phi_1} \tag{8.6}$$

$$E_2(x) = E_o e^{-\alpha d_2} e^{j\beta x} e^{j\phi_2} \tag{8.7}$$

Only the relative phase between ϕ_1 and ϕ_2 matters, so without loss of generality, ϕ_1 can be assumed to be zero. The focal E-field is given by the coherent summation of $E_1(x)$ and $E_2(x)$ at $x = 0$; that is,

$$E_{focus} = E_F = E_1(0) + E_2(0) \tag{8.8}$$

or, using the attenuation of 1 dB/cm for 6-cm thick fatty breast tissue, it follows that the electric field drops to a value of 0.707 relative to a normalized value of unity on the skin surface

$$E_F = 0.707 + 0.707e^{j\phi_2} \tag{8.9}$$

Taking the magnitude squared of (8.9) it follows that

$$\text{SAR}_{focus} = |E_F|^2 = 1 + \cos\phi_2 \tag{8.10}$$

Let E_{A1} and E_{A2} represent the electric field at surface point A due to plane waves 1 and 2, respectively. Similarly, let E_{B1} and E_{B2} represent the electric field at surface point B due to plane waves 1 and 2, respectively. The electric field on the opposing skin surfaces at points A and B is given by

$$E_A = E_{A1} + E_{A2} = 1.0 + 0.5e^{-j2.07}e^{j\phi_2} \tag{8.11}$$

and

$$E_B = E_{B1} + E_{B2} = 0.5e^{j2.07} + e^{j\phi_2} \tag{8.12}$$

Using the rounded value of 120° to approximate the 2.07 radians of phase shift in (8.11) and (8.12), it follows that

$$\text{SAR}_{LeftSkin} = \text{SAR}_A = |E_A|^2 = 1.25 + \cos(\phi_2 - 120°) \tag{8.13}$$

$$\text{SAR}_{RightSkin} = \text{SAR}_B = |E_B|^2 = 1.25 + \cos(\phi_2 + 120°) \tag{8.14}$$

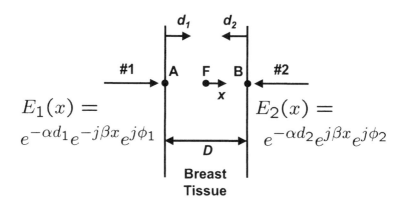

Figure 8.3 Diagram for coherent opposing plane waves irradiating lossy breast tissue of thickness D. In general, the two plane waves have the same oscillating frequency and different phase shifts ϕ_1 and ϕ_2.

The total SAR at the skin is equal to the sum of the SAR at the left and right skin surfaces; that is,

$$\text{SAR}_{TotalSkin} = \text{SAR}_A + \text{SAR}_B \tag{8.15}$$

or substituting (8.13) and (8.14) into (8.15) yields

$$\text{SAR}_{TotalSkin} = 2.5 + \cos(\phi_2 - 120°) + \cos(\phi_2 + 120°) \tag{8.16}$$

Using (8.10), the calculated SAR in a 6-cm thick breast tissue phantom at central depth (the desired focal position) as a function of the commanded phase ϕ_2 of applicator number 2 for opposing plane waves is shown in Figure 8.4. In Figure 8.4, the maximum value of SAR occurs when the commanded phase is zero degrees. When the commanded phase is equal to ±180°, the SAR is equal to zero (a null occurs at central depth). It is important to note that there is only one unique value of phase ϕ_2 that produces maximum SAR. In Figure 8.5, the SAR values computed from (8.13), (8.14), and (8.15) for the opposing skin surfaces versus commanded phase ϕ_2 are shown. For the sum of the SAR on the left and right skin surfaces, the minimum value of SAR occurs when the commanded phase is zero degrees. Thus, when the two-element adaptive phased array is focused at central depth, the sum of the SAR values for the opposing skin surfaces is a minimum. Equivalently, if a two-element adaptive phased array minimizes (attempts to null) the sum of the SAR on the opposing skin surfaces, a central deep focus is formed.

8.7 CLINICAL ADAPTIVE PHASED ARRAY THERMOTHERAPY SYSTEM FOR BREAST CANCER

A two-channel 915-MHz microwave adaptive phased array hyperthermia system (Microfocus APA 1000, Celsion (Canada) Limited, Columbia, Maryland) as shown in Figure 1.15 was developed for clinical trials for treating breast cancer in the intact breast. Each channel of the phased array contains an electronically variable microwave power amplifier (0 to 100W), an electronically variable phase shifter (0 to 360 degrees), and air-cooled linearly polarized rectangular waveguide applicators (Celsion (Canada) Limited, Model Numbers TEM-1 and TEM-2). The rectangular aperture dimensions of the pair of TEM-1 metallic waveguide applicators are 5.9 cm (parallel to the electric field) by 11.3 cm (perpendicular to the electric field). Similarly, the aperture dimensions of the pair of TEM-2 waveguide applicators are 6.5 cm by 13.0 cm. Clinical studies of single TEM applicators have been conducted for superficial cancers as described by Shidnia [73]. Dielectric loading of

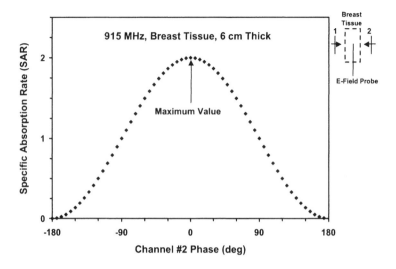

Figure 8.4 Calculated specific absorption rate in 6-cm thick breast tissue at central focal depth as a function of channel 2 phase for opposing coherent plane waves at 915 MHz.

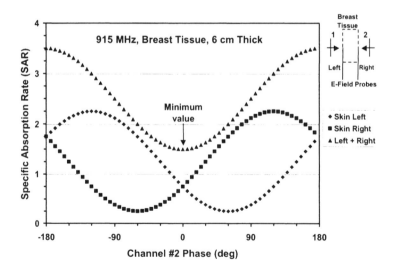

Figure 8.5 Calculated specific absorption rate on the skin surfaces of 6 cm thick breast tissue as a function of channel 2 phase for opposing coherent plane waves at 915 MHz.

the side walls of the rectangular waveguide region is used to obtain good impedance matching conditions and wide-field heating for a tissue load for the TEM waveguide applicator as described by Cheung [74] and Gauthrie [75]. Air cooling through the waveguide aperture is achieved by means of a fan mounted behind a perforated conducting screen that serves as a parallel reflecting ground plane for the input monopole feed for the waveguide. Taking into account the thickness of the dielectric slabs in contact with the waveguide side walls, the effective cross-sectional size for the air cooling is approximately 6.5 cm by 9.0 cm for the TEM-2 applicator. However, based on the dielectric parameter differences at 915 MHz between breast tumors and normal fatty (adipose-dominant) breast tissue, breast tumor is expected to heat more rapidly than normal fatty breast tissue, and the 50% SAR region for a tumor embedded in breast should tend to be concentrated primarily at or near the tumor.

As part of the focused microwave patient-treatment procedure, a 16-gauge (1.65-mm OD, 1.22-mm ID) closed-end plastic catheter with a metal introducer is inserted into the tumor under ultrasound guidance, and a single-use disposable combination E-field focusing sensor (1.12 mm OD) and fiber-optic temperature sensor is inserted in the catheter to focus the microwaves and measure the tumor temperature during thermotherapy. The combination sensor has the fiber-optic temperature sensor at the tip and the E-field sensor is 1.5 cm from the tip. Thermocouple probes (Physitemp Instruments, Inc.),

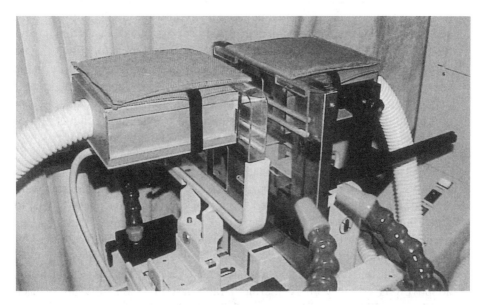

Figure 8.6 Photograph showing dual opposing TEM-2 915-MHz waveguide applicators, compression plates, and air-cooling equipment. (Photograph courtesy of Celsion (Canada) Limited.)

Type T copper-constantan, are used to measure the local temperature for the superficial tissues during treatment. The temperature at the focus is used as a real-time feedback signal during the treatment. This feedback signal is used to control the microwave output power level of the variable power amplifiers, which set and maintain the focal temperature in the tumor. The power and phase delivered to the two channels of the phased array are adjusted using digital-to-analog converters under computer control. The adaptively controlled microwave phase shift was determined by means of a gradient-search feedback algorithm as discussed earlier in this chapter. The microwave power level is adjusted to produce a rate of about 1°C temperature increase per minute until the desired feedback temperature is achieved.

8.8 CALCULATED SPECIFIC ABSORPTION RATE DISTRIBUTION FOR BREAST PHANTOM WITH AND WITHOUT TUMORS

To estimate the heating pattern in normal breast tissue and embedded tumor tissue that are exposed to focused microwave radiation, three-dimensional SAR patterns were calculated based on finite-difference time-domain (FDTD) theory as described in Chapter 3 with software developed by Taflove and colleagues at Northwestern University. In applying the FDTD method to thermotherapy treatments, the microwave applicators, tissues, and surrounding air are modeled by small cubes, and the electromagnetic field amplitude and phase at each cube is computed as a function of time based on Maxwell's electromagnetic field equations until steady-state fields are achieved. Referring to (8.4), once the steady-state electromagnetic field distribution has been computed, the SAR distribution is then calculated from the magnitude of the E-field squared multiplied by the ionic conductivity of each cube – variation in tissue density is ignored in this section.

As depicted in Figures 8.6 and 8.7, dual-opposing waveguide applicators (TEM-2 applicators, Celsion (Canada) Limited, Columbia, Maryland) operating at 915 MHz were modeled using the FDTD method. The applicators were simulated as coherently combined (common oscillator and phase tuned) to focus the radiated beams at the central position in 6-cm-thick homogeneous normal fatty breast tissue (dielectric constant 12.5 and ionic conductivity 0.21 S/m). The applicators are assumed to radiate through thin sheets of rigid acrylic material that simulate the plates used for breast compression in the adaptive phased array breast thermotherapy system. The side walls of each TEM-2 metallic waveguide are loaded with high dielectric constant material, which is used to shape the radiation inside the waveguide aperture. The waveguide applicators are linearly polarized with the alignment of the E-field

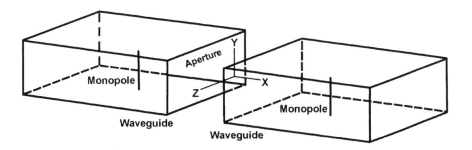

Figure 8.7 Geometry showing dual-opposing microwave phased array waveguide applicators. (From [29] Fenn, A.J., *Breast Cancer Treatment by Focused Microwave Thermotherapy*, 2007, Jones and Bartlett Publishers, Sudbury, MA, www.jbpub.com. Reprinted with permission.)

in the y direction as in Figure 8.7. As shown in Figure 8.2, in a typical focused microwave thermotherapy treatment, the E-field polarization is aligned with the direction from the nipple to the base (chest wall) of the breast. For this geometry, the focused microwave energy propagates in a direction approximately parallel to the chest wall. Little microwave energy would be expected to penetrate the chest wall because the microwave reflection coefficient at low incidence angles is very high. The metallic waveguide walls are modeled with a dielectric constant 1.0 and ionic conductivity 3.7×10^7 S/m. In the FDTD computer simulation, a 5-mm flat sheet of acrylic material (dielectric constant 2.6, ionic conductivity 0.00013 S/m) is adjacent to each applicator and parallel to the waveguide aperture. Between the two opposing TEM-2 applicators is a 6-cm-thick homogeneous normal fatty breast tissue phantom. The remaining volume is filled with 0.5-cm cubic cells that model air (dielectric constant 1.0, ionic conductivity 0.0 S/m).

The SAR distributions in the simulated compressed breast were calculated by squaring the electric-field amplitude and multiplying by the electrical conductivity of the simulated tissue (refer to (8.4)). SAR is often described in levels (50% is usually designated as the effective heating zone) relative to the maximum SAR value of 100%. The SAR is proportional to the initial rise in temperature per unit time, ignoring blood flow and thermal conduction effects. SAR patterns were computed in the xz-plane as shown in Figures 8.8 and 8.9 for homogeneous normal fatty breast tissue and homogeneous normal fatty breast tissue with two tumors, respectively. Figure 8.8 shows the top view (xz plane, $y = 0$) SAR pattern (75% and 50% SAR contours). The Figure 8.8 SAR pattern exhibits an elongated 50% SAR region. The result shown in Figure 8.8 indicates that a large volume of deep breast tissues can be heated by the adaptive phased array with focused TEM-2 waveguide applicators,

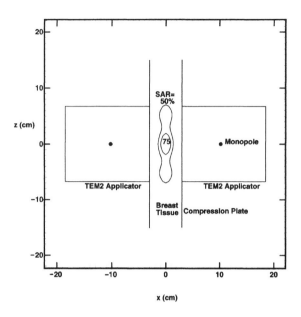

Figure 8.8 Calculated SAR contour pattern in the xz plane, $y = 0$, using the
FDTD method for dual-opposing focused microwave phased array
TEM-2 waveguide applicators. The applicators are focused at central
depth in simulated 6-cm thick homogeneous normal fatty breast tissue.
(From [29] Fenn, A.J., *Breast Cancer Treatment by Focused Microwave
Thermotherapy*, 2007, Jones and Bartlett Publishers, Sudbury, MA,
www.jbpub.com. Reprinted with permission.)

whereas the superficial tissues are not substantially heated. Any high-water-
content tissues exposed to this large heating field will be preferentially heated
compared to the surrounding normal fatty breast tissue as demonstrated in
the following paragraphs. To demonstrate selective (preferential) heating,
two spherically shaped 1.5-cm-diameter simulated tumors (dielectric constant
58.6, electrical conductivity 1.05 S/m for breast carcinoma as listed in Table
2.1) were embedded in the simulated normal fatty breast tissue phantom with
5-cm spacing on the z axis, and the FDTD SAR calculation for the top view
is shown in Figure 8.9. From a comparison of this result with Figure 8.8, it is
clear that the 50% SAR contour pattern has changed significantly and the two
high-water-content tumor regions are selectively heated.

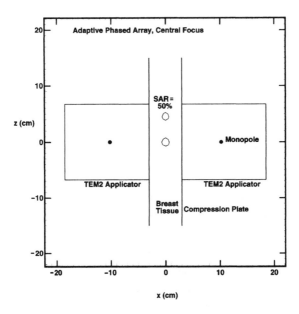

Figure 8.9 Calculated SAR contour pattern in the xz plane, $y = 0$, using the FDTD method for dual-opposing focused microwave phased array TEM-2 waveguide applicators. The applicators are focused at central depth in simulated 6-cm-thick homogeneous normal fatty breast tissue with two embedded 1.5 cm diameter spherically shaped breast tumors. (From [29] Fenn, A.J., *Breast Cancer Treatment by Focused Microwave Thermotherapy*, 2007, Jones and Bartlett Publishers, Sudbury, MA, www.jbpub.com. Reprinted with permission.)

8.9 MEASURED SPECIFIC ABSORPTION RATE DISTRIBUTION FOR BREAST PHANTOM WITH VARIABLE-SIZE TUMORS

To demonstrate selective tumor heating with a focused microwave phased array breast thermotherapy system, a simulated breast tumor mass of variable size was embedded in simulated normal fatty breast tissue and the SAR pattern was measured. Figure 8.6 shows a photograph of the externally focused adaptive phased array thermotherapy applicators with breast compression plates (3-mm-thick acrylic). Dual-opposing 915-MHz TEM-2 waveguide applicators (Celsion (Canada) Limited) were used in these experiments. To verify the wide-field aperture distribution of this type of thermotherapy applicator, an electric-field probe was manually scanned across the long dimension of the applicator aperture. The measured aperture electric field for a single TEM-2 applicator (without a tissue load) is shown in Figure 8.10. The measured -3 dB (50%) beamwidth is approximately 8.7 cm, which suggests that a wide heating region can be generated with this applicator as desired for treating breast cancer. As described below, a multislice breast phantom was used to simulate the breast for microwave-heating SAR experiments.

The breast phantom contained fatty dough material (approximately 66.7% flour, 30.0% oil, and 3.3% NaCl solution (0.9% NaCl per liter of water by weight), as described by Lagendijk [76], with microwave properties similar to normal fatty breast tissue. The breast tumor was simulated with high-water-content muscle phantom tissue (approximately 75.2% water, 1.0% NaCl, 15.4% Polyethylene powder, and 8.4% TX-151 gelling agent) described by Chou [77]. The 6-cm-thick breast phantom was constructed out of six slices using acrylic frames – each frame provided a 1-cm-thick slice simulating normal fatty breast tissue with a simulated 2-cm-thick cancerous breast tumor modeled in the two central frames. The slices were numbered 1 through 6, with slices 1 and 6 being the surface slices, and slices 3 and 4 the central slices. Five thermocouple (TC) catheter tracts spaced 1 cm apart (located between the slices) were used and had subsurface depths of 1, 2, 3, 2, and 1 cm, respectively. Thermocouple probes (Physitemp Instruments, Inc, Clifton, New Jersey) were inserted in the plastic catheters and were moved linearly using a computer-controlled probe scanning mechanism to 15 positions that were

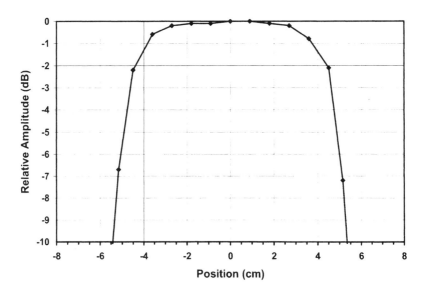

Figure 8.10 Measured electric-field aperture distribution for a single TEM-2 915-MHz waveguide applicator.

Acrylic Breast Phantom Frame

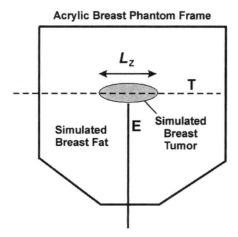

Figure 8.11 Drawing of a central breast phantom slice, 1-cm thick, with an E-field sensor catheter and a temperature sensor catheter track inserted through a simulated breast tumor. The maximum tumor dimension (denoted L_z) was varied from 2 to 8 cm. (From [29] Fenn, A.J., *Breast Cancer Treatment by Focused Microwave Thermotherapy*, 2007, Jones and Bartlett Publishers, Sudbury, MA, www.jbpub.com. Reprinted with permission.)

1 cm apart. A drawing of a central phantom slice is shown in Figure 8.11 with an example tumor modeled, having width L_z, at the approximate center of the slice. An E-field sensor probe (coaxial cable with the center pin extended) was placed with the tip at the approximate central depth of the tumor. A catheter for the thermocouple sensor was placed laterally in the tumor and was perpendicular to the E-field polarization. In Figure 8.11, the E-field probe catheter tract is denoted by the letter E and the thermocouple sensor catheter tract is denoted by the letter T. In each of three different experiments, the two middle sections (slices 3 and 4) contained an equal-size tumor phantom (1 cm × 2 cm × 2 cm, 1 cm × 2 cm × 6 cm, or 1 cm × 2 cm × 8 cm) so that when the two slices were placed together, central tumor masses with approximate elliptical dimensions ($L_x = 2, L_y = 2, L_z = 2$ cm), ($L_x = 2, L_y = 2, L_z = 6$ cm), or ($L_x = 2, L_y = 2L_z = 8$ cm) were modeled.

Heating was performed at 50 watts per channel over 20-second bursts with 20 minutes between measurements to allow the phantom temperature to return to equilibrium – a complete set of measurements for the 15 probe positions in the catheters took approximately 5 hours. The SAR was calculated by the rise in temperature at each measurement position over the 20-second heating interval (refer to (8.5)), and then was plotted as normalized SAR with

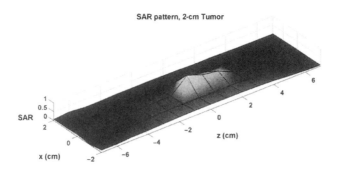

Figure 8.12 Measured SAR profiles for a simulated 6-cm compressed breast phantom with a $2 \times 2 \times 2$ cm simulated breast tumor heated by an adaptively focused 915-MHz microwave phased array. (From [29] Fenn, A.J., *Breast Cancer Treatment by Focused Microwave Thermotherapy*, 2007, Jones and Bartlett Publishers, Sudbury, MA, www.jbpub.com. Reprinted with permission.)

a peak value of 1.0. A normalized SAR value of 0.5 (50%) is typically used to estimate the effective heating zone. The microwave applicators are designed for clinical treatments so that a gap region (typically 1 cm or more) is provided between the applicator and the breast tissue. The gap region allows air to flow from external air tubes that are pointed into the gap to cool the region in proximity to the skin and base of each side of the breast and chest wall region. For this phantom testing procedure, air cooling was not used. The adaptive microwave phased array thermotherapy system was focused using the E-field probe sensor in the simulated tumor as the feedback signal. The measured SAR profile results for the three different tumor sizes (2, 6, and 8 cm width) are shown in Figures 8.12, 8.13, and 8.14. The SAR heating zone selectively expands in size according to the tumor size and encompasses nearly the entire tumor mass.

The measured data presented in this section, in which the adaptive focusing algorithm was used, show that the desired selective heating of a high-water-content simulated homogeneous tumor mass is achieved in a nonperfused breast phantom simulating normal fatty breast tissue. Based on these measured phantom data, the Microfocus APA-1000 focused microwave phased array thermotherapy system is capable of heating breast tumors that can vary in size from small to large diameter. In the next section, the results of adaptive phased array focused microwave thermotherapy trials for small and large animals are briefly discussed.

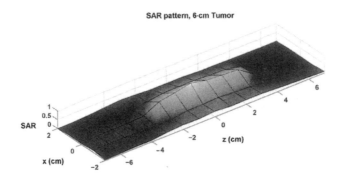

Figure 8.13 Measured SAR profiles for a simulated 6-cm compressed breast phantom with a $2 \times 2 \times 6$ cm simulated breast tumor heated by an adaptively focused 915-MHz microwave phased array. (From [29] Fenn, A.J., *Breast Cancer Treatment by Focused Microwave Thermotherapy*, 2007, Jones and Bartlett Publishers, Sudbury, MA, www.jbpub.com. Reprinted with permission.)

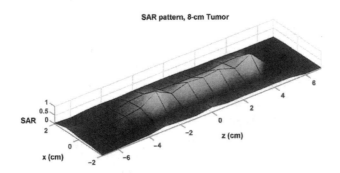

Figure 8.14 Measured SAR profiles for a simulated 6-cm compressed breast phantom with a $2 \times 2 \times 8$ cm simulated breast tumor heated by an adaptively focused 915-MHz microwave phased array. (From [29] Fenn, A.J., *Breast Cancer Treatment by Focused Microwave Thermotherapy*, 2007, Jones and Bartlett Publishers, Sudbury, MA, www.jbpub.com. Reprinted with permission.)

8.10 MEASUREMENTS OF THERMAL DISTRIBUTION IN ANIMALS

Thermotherapy testing of the adaptive phased array breast thermotherapy system with the adaptive focusing algorithm described in this chapter was conducted in small animals [24] from March 1997 to May 1998 at the Massachusetts General Hospital (MGH) Center for Imaging and

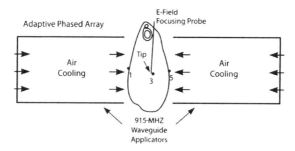

Figure 8.15 Geometry for animal trials test (rabbit hind leg) with an adaptive phased array of opposing waveguide applicators and an E-field focusing probe at a central tissue site. The temperature sensors are labeled 1, 3, 5. (From [24] with permission from Informa Healthcare, www.informaworld.com.)

Pharmaceutical Research (CIPR). The in vivo trials were conducted in several New Zealand White rabbits. The study was approved by the Subcommittee on Research Animal Care and conducted in an Association for Assessment and Accreditation of Laboratory Animal Care (AAALAC)-accredited facility.

The rabbit hind leg provides a model of perfused muscle tissue in the thigh region with a subcutaneous fat layer in the range of 2- to 3-mm thick. Muscle is significantly more difficult to penetrate with microwaves (915-MHz microwave loss = 2.9 dB/cm) compared to normal fatty breast tissue (915-MHz microwave loss = 1 dB/cm) – refer to Table 3.3. Thus, for demonstrating breast tumor heating in 4- to 8-cm-thick compressed breast tissue, these tests in microwave lossy muscle tissue can represent more difficult heating conditions compared to human trials. The animals were anesthetized with halothane prior to setup of the experiments and clinically monitored during the treatments. Experiments with adaptive microwave phased array thermotherapy TEM2 applicators opposing the hind leg were conducted in animals with and without tumors. All animals were euthanized via lethal injection following the treatments.

Figure 8.15 shows the animal trial test geometry for the dual-opposing adaptive phased array focused microwave applicators. Note that breast compression plates were not used in these experiments because compression of the rabbit hind leg was not required. The animal trials used thermocouple temperature probes (Physitemp Instruments, Inc., Clifton, New Jersey) - these probes are denoted 1, 3, and 5 in Figure 8.15. The rabbit hind leg thickness was approximately 4 cm and the spacing between the temperature probes for this test was approximately 2 cm. Temperature probes 1 and 5 were positioned on the opposing skin surfaces of the hind leg. The hind leg was shaved so that the skin-surface-mounted thermocouple probes could be taped to the skin and could contact and measure the temperatures of the skin. The E-field probe and temperature probe 3 were inserted into the tissue within closed-end plastic catheters. Probe 3 was at the focal depth of 2 cm. The tips of all probes were approximately aligned with the projected midpoint of the waveguide apertures. The temperature probes were aligned approximately perpendicular to the electric-field polarization of the 915-MHz waveguide applicators so that they would not scatter significant amounts of microwave energy during treatments or not be significantly heated by the microwave field. To focus the dual-applicator phased array thermotherapy system adaptively, the E-field probe was inserted into the central tissue position (2 cm depth) approximately parallel to and strongly coupled with the electric field radiated by the applicators.

For the dual opposing adaptively focused microwave phased array, the 915-MHz electronic phase shifters were adjusted typically in a total of 5 to 10 seconds using the adaptive phased array algorithm with acceleration, as described in Section 8.5, to focus the microwave radiation rapidly at the central tissue site. The locations of all probes were verified using an X-ray computerized tomography (CT) scan prior to each treatment. Note that in clinical practice, the orientation of the E-field probe and temperature probe in the breast is verified using ultrasound imaging. The two 915-MHz microwave applicators used air cooling (room temperature air, 20°C) to limit the amount of heating induced in the skin. For these tests, the target feedback temperature of centrally located temperature probe 3 was 43°C for 60-minute treatments or 46°C for shorter-duration treatments [24].

An example experiment with the adaptive phased array using the target feedback temperature of 46°C at the tumor for 8 minutes was conducted. Figure 8.16 shows the measured temperature versus time data and demonstrates that the desired feedback temperature is achieved after approximately 8 minutes of heating and is then maintained for the desired 8-minute time interval. Referring to (1.1), the tumor equivalent thermal dose [78] delivered was 91.7 minutes relative to 43°C. During the entire treatment interval, the skin temperatures were below 35°C. During this treatment, the average microwave power per applicator was in the range of 25 to 30W [24].

In these preclinical tests of a Microfocus APA-1000 dual-opposing air-cooled adaptive microwave phased array thermotherapy system with adaptive focusing algorithm, a feedback temperature of either 43°C or 46°C can be achieved at a central tissue depth in a rabbit hind leg model without excessive superficial heating. The adaptive phased array experiment described here used two coherent microwave channels containing electronically adjustable phase shifters to focus the microwave energy at an E-field feedback probe. This adaptive phased array system has a significant advantage over a nonadaptive phased array. A nonadaptive phased array with two channels could, in theory, produce a null, a maximum, or an intermediate value of E-field depending on whether the two radiated waves are 180 degrees out-of-phase, completely in-phase, or partly out-of-phase, respectively. Because the adaptive phased array automatically focuses the E-field in the presence of all tissue-scattering structures, this type of array should provide more reliable deep-focused heating compared to manually adjusted or pretreatment planning-controlled phased arrays.

The ability of a dual-applicator adaptive microwave phased array to heat semi-deep tissues was confirmed in a small live animal model. The experiments were performed in high-water-content perfused muscle tissue that theoretically is more difficult to heat at depth compared to the same thickness of low-water-content fatty breast tissue. Based on the measured data from these experiments in a rabbit hind leg, the adaptive microwave phased array breast thermotherapy system demonstrated the necessary safety to warrant additional investigations in humans. However, prior to the start of human trials described in the next chapter, additional tests in larger animals were conducted as briefly described below.

Additional preclinical safety assessments of the heating capabilities of the two-channel 915-MHz Microfocus APA 1000 hyperthermia system with adaptive focusing algorithm were performed in the hind legs of pigs, which are significantly thicker (50% or greater) than the rabbit hind legs described above. In addition, the large-animal study included the use of dielectric plates

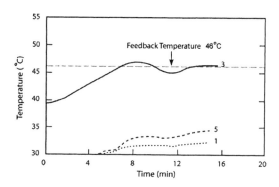

Figure 8.16 Focused microwave thermotherapy treatment of rabbit hind leg using an adaptive phased array thermotherapy system. The surface sensors are labled 1 and 5 and the central probe at the focus is probe 3. The feedback temperature is 46°C. (From [24] with permission from Informa Healthcare, www.informaworld.com.)

to compress the tissue, whereas the rabbit study did not utilize compression plates. Focused microwave thermotherapy tests in large animals (pigs) were conducted in March to April 1998 at Oxford University in Oxford, England, in collaboration with Hammersmith Hospital in London as decribed by Gavrilov [25]. Female English Large White pigs, aged 7 to 8 months and weighing 50 to 70 kg, were used in these preclinical tests that were conducted under veterinarian care. Prior to microwave heating, the animals were anaesthetized using a halothane (2 to 3%), nitrous oxide (30%), and oxygen mixture and were intubated. They were positioned supinely during thermotherapy sessions. The site treated in each animal was the muscle region of the hind leg. The leg was supported in a near vertical position. The dimensions of the legs (variation and range of muscle thickness, bone thickness) were measured and the borders of a bone-free volume of tissue were estimated to determine an appropriate treatment volume. The results of these preclinical experiments using the Microfocus APA 1000 focused microwave thermotherapy system showed that when the muscle region of the hind legs of pigs can be compressed to 6.5 to 7 cm, a pair of parallel opposed, coherently driven 915-MHz applicators could elevate the temperature in the central tissue to therapeutic levels (43°C) without overheating superficial tissues. In these tests, the phase difference between applicators was determined and controlled by the adaptive phase focusing algorithm described in this chapter. The results obtained supported the potential of using the adaptive microwave phased array technique for thermotherapy treatment of centrally located breast tumors in a 4- to 8-cm-thick compressed breast, particularly because 915-MHz microwave attenuation is about 1 dB/cm for normal breast tissue and about 3 dB/cm in muscle corresponding to the pig model. The results of the large-animal study demonstrated further safety for human clinical trials for breast cancer patients.

8.11 SUMMARY

This chapter has described an adaptive phased array focused microwave thermotherapy system for treating breast cancer in the intact breast. Surgical considerations for breast cancer were discussed in Section 8.2. Reducing positive tumor margins is critical in reducing the recurrence rate of breast cancer. Microwave considerations for breast tissue and breast cancer were discussed in Section 8.3. Breast cancer tumors can be heated rapidly with microwaves while lower-water, lower-ion content normal fatty breast tissue would heat less. A detailed description of the adaptive phased array thermotherapy system concept for breast cancer was given in Section 8.4. In Section 8.5, the mathematical formulation for an adaptive phased array algorithm for a two-channel focused microwave thermotherapy system was given. Ray-tracing calculations for the specific absorption rate (SAR) delivered to breast tissue by a 915-MHz focused microwave phased array were presented in Section 8.6. In Section 8.7 a clinical adaptive phased array focused microwave thermotherapy system for treating breast cancer in the intact breast was described. In Section 8.8, the finite-difference time-domain technique was used to compute the SAR delivered by the focused microwave waveguide phased array for breast tissue alone and breast tissue with tumors. The FDTD simulations clearly show preferential heating of simulated breast cancer tumors embedded in simulated normal fatty breast tissue. The measured SAR for a compressed breast phantom with variable size phantom tumors heated by the adaptive focused microwave phased array thermotherapy system was given in Section 8.9. The measurements showed that simulated phantom tumors with variable size are preferentially heated compared to the surrounding simulated normal fatty breast phantom material. Animal trial testing with an adaptive microwave phased array thermotherapy system was briefly described in Section 8.10, and it was demonstrated that a central tissue site could be heated to therapeutic temperatures while maintaining lower temperature at the surface tissue. The next chapter describes an example clinical application for adaptive phased array focused microwave thermotherapy in combination with chemotherapy for preoperative treatment of large breast carcinomas.

References

[1] American Cancer Society, Cancer Facts & Figures 2008, Atlanta: *American Cancer Society, Inc.*, p. 4.

[2] American Cancer Society, Breast Cancer Facts & Figures 2007-2008, Atlanta: *American Cancer Society, Inc.*, p. 1.

[3] Newman, L.A., and H.M. Kuerer, "Advances in Breast Conservation Therapy," *J Clinical Oncology*, Vol. 23, No. 8, 2005, pp. 1685-1697.

[4] Carlson, R.W., et al., "The NCCN Invasive Breast Cancer Clinical Practice Guidelines in Oncology," *J Nat Comprehensive Cancer Network*, Vol. 5, No. 3, 2007, pp. 246-312.

[5] Singletary, S.E., "Surgical Margins in Patients with Early-Stage Breast Cancer Treated with Breast Conservation Therapy," *American Journal of Surgery*, Vol. 184, 2002, pp. 383-393.

[6] Fisher, B. et al., "Twenty-Year Follow-Up of a Randomized Trial Comparing Total Mastectomy, Lumpectomy, and Lumpectomy Plus Irradiation for the Treatment of Invasive Breast Cancer," *New England J Medicine*, Vol. 347, 2002, pp. 1233-1241.

[7] Katipamula, R., et al., "Trends in Mastectomy Rates at the Mayo Clinic Rochester: Effects of Surgical Year and Preoperative MRI," 2008 American Society of Clinical

Oncology 44th Annual Meeting, May 30-June 3, 2008, Chicago, IL, *Journal of Clinical Oncology*, Supplement, Vol. 26, No. 15S, Part I of II, Abstract 509, 2008, p. 9s.

[8] Fenn, A.J., "Adaptive Nulling Hyperthermia Array," US Patent No. 5,251,645, October 12, 1993.

[9] Fenn, A.J., "Adaptive Focusing and Nulling Hyperthermia Annular and Monopole Phased Array Applicators," US Patent No. 5,441,532, August 15, 1995.

[10] Fenn, A.J., "Adaptive Hyperthermia for Improved Thermal Dose Distribution," In: *Radiation Research: A Twentieth Century Perspective,* Vol. 1 (Congress Abstracts), Chapman J.D., W.C. Dewey, G.F. Whitmore, (eds.), San Diego, Calif.: Academic Press, 1991, p. 290.

[11] Fenn, A.J., and G.A. King, "Adaptive Nulling in the Hyperthermia Treatment of Cancer," *The Lincoln Laboratory Journal,* Lincoln Laboratory, Massachusetts Institute of Technology, Vol. 5, No. 2, 1992, pp. 223-240.

[12] Fenn, A.J., B.A. Bornstein, G.K. Svensson, and H.F. Bowman, "Minimally Invasive Monopole Phased Arrays for Hyperthermia Treatment of Breast Carcinomas: Design and Phantom Tests," *International Symposium on Electromagnetic Compatibility,* Sendai, Japan: 17-19 May 1994, pp. 566-569.

[13] Fenn, A.J., C.J. Diederich, and P.R. Stauffer, "An Adaptive-Focusing Algorithm for a Microwave Planar Phased-Array Hyperthermia System," *The Lincoln Laboratory Journal,* Lincoln Laboratory, Massachusetts Institute of Technology, Vol. 6, No. 2, 1993, pp. 269-288.

[14] Fenn, A.J., and G.A. King, "Experimental Investigation of an Adaptive Feedback Algorithm for Hot Spot Reduction in Radio-Frequency Phased-Array Hyperthermia," *IEEE Trans Biomed Eng.,* Vol. 43, No. 3, 1994, pp. 273-280.

[15] Fenn, A.J., and G.A. King, "Adaptive Radio Frequency Hyperthermia Phased Array System for Improved Cancer Therapy: Phantom Target Measurements," *Int J Hyperthermia*, Vol. 10, No. 2, 1994, pp. 189-208.

[16] Fenn, A.J., A.Y. Cheung, and H. Cao, "Adaptive Focusing Experiments for Minimally Invasive Monopole Phased Arrays in Hyperthermia Treatment of Breast Cancer," *16th Annual IEEE Engineering in Medicine and Biology Society Int Conf,* Baltimore, Maryland: November 3-6, 1994, pp. 766-767.

[17] Fenn, A.J., "Minimally Invasive Monopole Phased Arrays for Hyperthermia Treatment of Breast Cancer," In: *Proc. 1994 Int. Symp. on Antennas,* Nice, France: November 8-10, 1994, pp. 418-421.

[18] Fenn, A.J., "Minimally Invasive Monopole Phased Array Hyperthermia Applicators and Method for Treating Breast Carcinomas," US Patent No. 5,540,737, July 30, 1996.

[19] Fenn, A.J., "Adaptive Focusing Experiments with an Air-Cooled 915-MHz Hyperthermia Phased Array for Deep Heating of Breast Carcinomas," *Proc of the Surgical Applications of Energy Sources Conference,* Estes Park, Colorado, May 17-19, 1996.

[20] Fenn, A.J. and G.A. King, "Experimental Investigation of an Adaptive Feedback Algorithm for Hot Spot Reduction in Radio-Frequency Phased Array Hyperthermia," *IEEE Trans Biomedical Eng,* Vol. 43, No. 3, 1996, pp. 273-280.

[21] Fenn, A.J., "Thermodynamic Adaptive Phased Array System for Activating Thermosensitive Liposomes in Targeted Drug Delivery," US Patent No. 5,810,888, September 22, 1998.

[22] Sathiaseelan, V., A.J. Fenn, and A. Taflove, "Recent Advances in External Electromagnetic Hyperthermia," In: Chapter 10 of *Advances in Radiation Treatment*, Mittal, B.B., J.A. Purdy, and K.K. Ang, (eds.), Boston, Massachusetts: Kluwer Academic Publishers, 1998, pp. 213-245.

[23] Fenn, A.J., V. Sathiaseelan, G.A. King, and P.R. Stauffer, "Improved Localization of Energy Deposition in Adaptive Phased Array Hyperthermia Treatment of Cancer," *J Oncol Management*, Vol. 7, No. 2, 1998, pp. 22-29.

[24] Fenn, A.J., G.L. Wolf, and R.M. Fogle, "An Adaptive Phased Array for "Targeted Heating of Deep Tumors in Intact Breast: Animal Study Results," *Int J Hyperthermia*, Vol. 15, No. 1, 1999, pp. 45-61.

[25] Gavrilov, L.R., J.W. Hand, J.W. Hopewell, and A.J. Fenn, "Pre-clinical Evaluation of a Two-Channel Microwave Hyperthermia System with Adaptive Phase Control in a Large Animal," *Int J Hyperthermia*, Vol. 15, No. 6, 1999, pp. 495-507.

[26] Gardner, R.A., H.I. Vargas, J.B. Block, C.L. Vogel, A.J. Fenn, G.V. Kuehl, and M. Doval, "Focused Microwave Phased Array Thermotherapy for Primary Breast Cancer," *Ann Surg Oncol*, Vol. 9, No. 4, 2002, pp. 326-332.

[27] Vargas, H.I., W.C. Dooley, R.A. Gardner, K.D. Gonzalez, S.H. Heywang-Kobrunner, and A.J. Fenn, "Success of Sentinel Lymph Node Mapping After Breast Cancer Ablation with Focused Microwave Phased Array Thermotherapy," *Am J Surg*, Vol. 186, 2003, pp. 330-332.

[28] Vargas, H.I., W.C. Dooley, R.A. Gardner, K.D. Gonzalez, S.H. Heywang-Kobrunner, and A.J. Fenn, "Focused Microwave Phased Array Thermotherapy for Ablation of Early-Stage Breast Cancer: Results of Thermal Dose Escalation," *Ann Surg Oncol*, Vol. 11, No. 2, 2004, pp. 139-146.

[29] Fenn, A.J., *Breast Cancer Treatment by Focused Microwave Thermotherapy*, Sudbury, MA: Jones and Bartlett, 2007.

[30] Vargas, H.I., W.C. Dooley, A.J. Fenn, M.B. Tomaselli, and J.K. Harness, "Study of Preoperative Focused Microwave Phased Array Thermotherapy in Combination with Neoadjuvant Anthracycline-Based Chemotherapy for Large Breast Carcinomas," *Cancer Therapy*, Vol. 5, 2007, published online (www.cancer-therapy.org), November 25, 2007, pp. 401-408.

[31] Dooley, W.C., H.I. Vargas, A.J. Fenn, M.B. Tomaselli, and J.K. Harness, "Randomized Study of Preoperative Focused Microwave Phased Array Thermotherapy for Early-Stage Invasive Breast Cancer," *Cancer Therapy*, Vol. 6, 2008, published online (www.cancer-therapy.org), August 25, 2008, pp. 395-408.

[32] Vargas, H.I., W.C. Dooley, A.J. Fenn, M.B. Tomaselli, and J.K. Harness, "A Protocol for Focused Microwave Thermotherapy in Combination with Neoadjuvant Anthracycline-Based Chemotherapy for Investigating Tumor Response of Large Invasive Breast Carcinomas," *American J Clin Oncol*, Abstract, Vol. 31, No. 5, October, 2008.

[33] Haagensen, C.D., *Diseases of the Breast,* 3rd ed., Philadelphia, Pa: WB Saunders, 1986, pp. 8-46.

[34] Arthur, D.W., and F.A. Vicini, "Accelerated Partial Breast Irradiation as a Part of Breast Conservation Therapy," *J Clin Oncology,* Vol. 23, No. 8, March 10, 2005, pp. 1726-1735.

[35] Hamilton, A., and G. Hortobagyi, "Chemotherapy: What Progress in the Last 5 Years?" *J Clin Oncol,* Vol. 23, 2005, pp. 1760-1775.

[36] Kaufmann, M., et al., "Recommendations from an International Expert Panel on the Use of Neoadjuvant (Primary) Systemic Treatment of Operable Breast Cancer, An Update," *J Clin Oncol,* Vol. 24, No. 12, 2006, pp. 1940-1949.

[37] Kaufmann, P., C.E. Dauphine, M.P. Vargas, M.L. Burla, N.M. Isaac, K.D. Gonzalez, D. Rosing, and H.I. Vargas, "Success of Neoadjuvant Chemotherapy in Conversion of Mastectomy to Breast Conservation Surgery," *Am Surgeon,* Vol. 72, No. 10, 2006, pp. 935-938.

[38] Fisher, B., et al., "Effect of Preoperative Chemotherapy on Local-Regional Disease in Women with Operable Breast Cancer: Findings from the National Surgical Adjuvant Breast and Bowel Project B-18," *J Clin Oncol,* Vol. 15, No. 7, 1997, pp. 2483-2493.

[39] Fisher, B., et al., "Effect of Preoperative Chemotherapy in the Outcome of Women with Operable Breast Cancer," *J Clin Oncol,* Vol. 16, No. 8, 1998, pp. 2672-2685.

[40] Wolmark, N., J. Wang, E. Mamounas, J. Bryant, and B. Fisher, "Preoperative Chemotherapy in Patients with Operable Breast Cancer: Nine-Year Results from National Surgical Adjuvant Breast and Bowel Project B-18," *J Natl Cancer Inst Monographs,* No. 30, 2001, pp. 96-102.

[41] Hall-Craggs, M.A., "Interventional MRI of the Breast: Minimally Invasive Therapy," *Eur Radiol,* Vol. 10, No. 1, 2000, pp. 59-62.

[42] Singletary, S.E., "Minimally Invasive Techniques in Breast Cancer Treatment," *Seminars in Surgical Oncology,* Vol. 20, 2001, 246-250.

[43] Noguchi, M., "Minimally Invasive Surgery for Small Breast Cancer," *J Surg Oncol,* Vol. 84, 2003, pp. 94-101.

[44] Agnese, D.M., and W.E. Burak, "Ablative Approaches to the Minimally Invasive Treatment of Breast Cancer," *Cancer Journal,* Vol. 11, 2005, pp. 77-82.

[45] Huston, TL, and R.M. Simmons, "Ablative Therapies for the Treatment of Malignant Diseases of the Breast," *Am J Surg,* Vol. 189, 2005, pp. 694-701.

[46] van Esser, S., M.A.A.J. van den Bosch, P.J. van Diest, W.Th.M. Mali, I.H.M. Borel Rinkes, R. van Hillegersberg, "Minimally Invasive Ablative Therapies for Invasive Breast Carcinomas: An Overview of Current Literature," *World J Surgery,* 2007, published online October 24, 2007.

[47] Jeffrey, S.S., R.L. Birdwell, D.M. Ikeda, et al., "Radiofrequency Ablation of Breast Cancer: First Report of an Emerging Technology," *Arch Surg,* Vol. 134, No. 10, 1999, pp. 1064-1068.

[48] Izzo, F., et al., "Radiofrequency Ablation in Patients with Primary Breast Carcinoma. A

Pilot Study of 26 Patients," *Cancer*, Vol. 92, No. 8, 2001, pp. 2036-2044.

[49] Singletary, S.E., B.D. Fornage, N. Sneige, et al., "Radiofrequency Ablation of Early-Stage Invasive Breast Tumors: an Overview," *Cancer J*, Vol. 8, 2002, pp. 177-180.

[50] Burak, W.E., D.M. Agnese, S.P. Povoski, et al., (2003) "Radiofrequency Ablation of Invasive Breast Carcinoma Followed by Delayed Surgical Excision," *Cancer*, Vol. 98, 2003, pp. 1369-1376.

[51] Fornage, B.D., et al., "Small ($\leq$2-cm) Breast Cancer Treated with US-Guided Radiofrequency Ablation: Feasibility Study," *Radiology*, Vol. 231, No. 1, 2004, pp. 215-226.

[52] Ross, M.I., and B.D. Fornage,"Radiofrequency Ablation of Early-Stage Breast Cancer," In: Chapter 9 of *Radiofrequency Ablation for Cancer: Current Indications, Techniques, and Outcomes,* Ellis, L.M., S.A. Curley, and K.K. Tanabe, (eds.), New York: Springer-Verlag, 2004, pp. 137-158.

[53] Dowlatshashi, K., D.S. Francescatti, K.J. Bloom, "Laser Therapy for Small Breast Cancers," *Am J Surgery*, Vol. 184, 2002, pp. 359-363.

[54] Dowlatshashi, K., J.J. Dieschbourg, K.J. Bloom, "Laser Therapy of Breast Cancer with 3-Year Follow-Up," *The Breast Journal*, Vol. 10, No. 3, 2004, pp. 240-243.

[55] Huber, P.E., et al., "A New Noninvasive Approach in Breast Cancer Therapy Using Magnetic Resonance Imaging-Guided Focused Ultrasound Surgery," *Cancer Res*, Vol. 61, 2001, pp. 8441-8447.

[56] Wu, F., et al., "Extracorporeal High Intensity Focused Ultrasound Treatment for Patients with Breast Cancer," *Breast Cancer Res Treat*, Vol. 92, 2005, pp. 51-60.

[57] Lu, X-Q., E.C. Burdette, B.A. Bornstein, J.L. Hansen, and G.K. Svensson, "Design of an Ultrasonic Therapy System for Breast Cancer Treatment," *Int Journal of Hyperthermia*, Vol. 12, No. 3, 1996, pp. 375-399.

[58] Hynynen, K., "Focused Ultrasound Surgery Guided by MRI," *Science & Medicine*, Vol. 3, No. 5, 1996, pp. 62-71.

[59] Pfleiderer, S.O., M.G. Freesmeyer, C. Marx, R. Kuhne-Heid, A. Schneider, and W.A. Kaiser, "Cryotherapy of Breast Cancer Under Ultrasound Guidance: Initial Results and Limitations," *Eur Radiol,* Vol. 12, 2002, pp. 3009-3014.

[60] Sabel, M.S., C.S. Kaufman, P. Whitworth, H. Chang, L.H. Stocks, R. Simmons, and M. Schultz, "Cryoablation of Early-Stage Breast Cancer: Work-in-Progress Report of a Multi-Institutional Trial," *Ann Surg Oncol,* Vol. 11, No. 5, 2004, pp. 542-549.

[61] Chaudhary, S.S., R.K. Mishra, A. Swarup, and J.M. Thomas, "Dielectric Properties of Normal and Malignant Human Breast Tissue at Radiowave and Microwave Frequencies," *Indian Journal of Biochemistry and Biophysics,* Vol. 21, 1984, pp. 76-79.

[62] Joines, W.T., Y. Zhang, C. Li, and R.L. Jirtle, "The Measured Electrical Properties of Normal and Malignant Human Tissues From 50 to 900 MHz," *Med Phys,* Vol. 21, No. 4, 1994, pp. 547-550.

[63] Surowiec, A.J., S.S. Stuchly, J.R. Barr, and A. Swarup, "Dielectric Properties of Breast

Carcinoma and the Surrounding Tissues," *IEEE Trans Biomed Eng,* Vol. 35, No. 4, 1988, pp. 257-263.

[64] Campbell, A.M., and D.V. Land, "Dielectric Properties of Female Human Breast Tissue Measured in Vitro at 3.2 GHz," *Phys Med Biol,* Vol. 37, No. 1, 1992, pp. 193-210.

[65] Burdette, E.C. "Electromagnetic and Acoustical Properties of Tissue," In *Physical Aspects of Hyperthermia,* Nussbaum, G.H., (ed.), AAPM Medical Physics Monographs, No. 8, 1982, pp. 105-130.

[66] Lazebnik, M., et al., "A Large-Scale Study of the Ultrawideband Microwave Dielectric Properties of Normal, Benign, and Malignant Breast Tissues Obtained from Cancer Surgeries," *Phys Med Biol,* Vol. 52, 2007, pp. 6093-6115.

[67] Lazebnik, M., et al., "A Large-Scale Study of the Ultrawideband Microwave Dielectric Properties of Normal Breast Tissue Obtained from Reduction Surgeries," *Phys Med Biol,* Vol. 52, 2007, pp. 2637-2656.

[68] Sha L., E.R. Ward, B. Story, "A Review of Dielectric Properties of Normal and Malignant Breast Tissue," *Proc IEEE SoutheastCon,* 2002, pp. 457-462.

[69] Gabriel, S., R.W. Lau, and C. Gabriel, "The Dielectric Properties of Biological Tissues: Part III. Parametric Models for the Dielectric Spectrum of Tissues," *Phys Med Biol,* Vol. 41, 1996, pp. 2271-2293. online (http://niremf.ifac.cnr.it/tissprop).

[70] Klein, R., et al., "Determination of Average Glandular Dose with Modern Mammography Units for Two Large Groups of Patients," *Phys Med Biol,* Vol. 42, 1997, pp. 651-671.

[71] Bassett, L., et al., "Stereotactic Core-Needle Biopsy of the Breast: A Report of the Joint Task Force of the American College of Radiology, American College of Surgeons, and College of American Pathologists," *CA, A Cancer Journal for Clinicians,* Vol. 47, 1997, pp. 171-190.

[72] Sullivan, D.C., C.A. Beam, S.M. Goodman, and D.L. Watt, "Measurement of Force Applied During Mammography," *Radiology,* Vol. 181, 1991, pp. 355-357.

[73] Shidnia, H., N. Hornback, G. Ford, and R.N. Shen, "Clinical Experience with Hyperthermia in Conjunction with Radiation Therapy," *Oncology,* Vol. 50, 1993, pp. 353-361.

[74] Cheung, A.Y., T. Dao, and J.E. Robinson, "Dual-Beam TEM Applicator for Direct-Contact Heating of Dielectrically Encapsulated Malignant Mouse Tumor," *Radio Science,* Vol. 12, No. 6(S) Supplement, 1977, pp. 81-85.

[75] Gautherie, M., (ed.), *Methods of External Hyperthermic Heating,* New York: Springer-Verlag, 1990, p. 33.

[76] Lagendijk, J.J.W., and P. Nilsson, "Hyperthermia Dough: a Fat and Bone Equivalent Phantom to Test Microwave/Radiofrequency Hyperthermia Heating Systems," *Phys Med Biol,* Vol. 30, No. 7, 1985, pp. 709-712.

[77] Chou, C.K., G.W. Chen, A.W. Guy, and K.H. Luk, "Formulas for Preparing Phantom Muscle Tissue at Various Radiofrequencies," *Biolectromagnetics,* Vol. 5, 1984, pp. 435-441.

[78] Sapareto, S.A., and W.C. Dewey, "Thermal Dose Determination in Cancer Therapy," *Int J Rad Oncol Biol Phys,* Vol. 10, 1984, pp. 787-800.

9

Adaptive Array for Breast Cancer: Clinical Results

9.1 INTRODUCTION

Ablation as part of a multimodality approach in the treatment of breast cancer is the subject of recent interest as described by Hall-Craggs [1], Singletary [2], Noguchi [3], Agnese and Burak [4], Huston and Simmons [5], and van Esser [6]. The use of thermal energy with radiofrequency [7-12], interstitial laser photocoagulation [13, 14], focused ultrasound [15-17], cryotherapy [18, 19], or adaptive phased array focused microwave thermotherapy [20-30] as described in Chapter 8 has demonstrated some success in achieving ablation of breast cancer tumors. For breast cancer thermotherapy, microwave energy is promising because it preferentially heats and damages high-water high-ion content breast carcinomas, compared to lesser degrees of heating that occurs in lower-water, lower-ion content normal fatty breast tissues [31-34]. Focused microwave thermotherapy has been explored as a preoperative treatment modality for breast cancer, to cause tumor cell kill as well as to cause tumor volume reduction as described by Gardner [20], Vargas [22, 29], Fenn [23], and Dooley [30]. For example, tumor cell kill and tumor volume reduction can potentially improve tumor margins as well as improve breast conservation. In cancer treatment using thermotherapy, referring to (1.1) and Table 1.2, an equivalent thermal dose in minutes of effective treatment relative to treatment at a temperature of 43°C (expressed as CEM(43°C)) is often used to quantify the treatment thermal dose [35].

The concept of a focused microwave thermotherapy system for treatment of breast cancer was shown in Figure 8.1 and a clinical system was shown in Figure 1.15. Four clinical studies [23] of adaptive phased array focused microwave thermotherapy for breast cancer have been conducted to date, as summarized in Table 9.1. In a Phase I safety study described by Gardner [20] conducted from December 1999 to July 2000, 10 patients with breast cancer in the intact breast who were scheduled for mastectomy received one treatment of focused microwave thermotherapy prior to receiving mastectomy. Breast cancer tumors ranged in maximum clinical size from 0.9 to 8 cm (mean 4.2 cm). The measured tumor thermal dose ranged from 24.5 to 100 equivalent thermal minutes relative to treatment at $43°C$. Eight of the 10 patients had a partial tumor response quantified by tumor shrinkage based on ultrasound measurements or had partial tumor cell kill based on pathology. In a Phase II dose escalation study described by Vargas [22] conducted from May 2001 to July 2002, 25 patients with invasive breast cancer in the intact breast received treatment of focused microwave thermotherapy prior to breast conserving surgery (wide-excision lumpectomy). Tumor size ranged from 0.7 to 2.5 cm (mean 1.8 cm) based on ultrasound measurements. In this dose-escalation study, based on data for 19 patients, a tumor cumulative thermal dose of 210 equivalent thermal minutes, relative to treatment at $43°C$, is predictive of 100% tumor cell kill (ablation).

Subsequently, two small Phase III randomized clinical studies of preoperative focused microwave thermotherapy were conducted as a heat-alone ablation treatment for patients with early-stage breast cancer, and as a heat treatment in combination with chemotherapy for patients with large breast cancer tumors. In the Phase III study of heat-alone focused microwave thermotherapy prior to breast conservation surgery [30] conducted from November 2002 to May 2004, 92 patients with early-stage breast cancer were enrolled. Of 34 patients that received focused microwave thermotherapy prior to surgery, none (0%) had positive tumor margins, and of 41 patients that received surgery alone, 4 (9.8%) had positive tumor margins ($P = 0.13$, not significant). In the Phase III study for patients with large breast cancer tumors [29] conducted from November 2002 to May 2004, 34 patients were enrolled. Median tumor volume reduction based on ultrasound measured tumor volume was 88.4% ($n=14$) in the preoperative thermochemotherapy arm compared to 58.8% ($n=10$) in the preoperative chemotherapy-alone control arm ($P = 0.048$, statistically significant). This study is the detailed subject matter of this chapter as a follow-on larger clinical study of focused microwave thermochemotherapy for patients with large breast cancer tumors is planned.

For patients with large primary breast carcinomas, the probable outcome is dependent on the stage at diagnostic presentation. In addition to low survival rates, patients with large breast cancer tumors may require mastectomy. Various regimens of chemotherapy are administered to breast cancer patients

Table 9.1

Clinical Trials of Adaptive Phased Array Thermotherapy for Breast Cancer

Study Type	Number of Patients Enrolled	Protocol
Phase I (Safety)	10	TT prior to mastectomy [20]
Phase II (Dose escalation)	25	TT prior to BCS [22]
Phase III (randomized)	92	TT prior to BCS [30]
Phase III (randomized)	34	TCT prior to breast surgery [29]

TT=thermotherapy, TCT=thermochemotherapy, BCS=breast conservation surgery.

in order to treat systemic cancer in an attempt to provide improved disease-free survival as well as overall survival [36, 37]. Recent treatment approaches for patients with large breast cancer tumors have explored chemotherapy regimens in a preoperative (neoadjuvant) setting. The impact on survival for primary preoperative AC (doxorubicin (an anthracycline) and cyclophosphamide) chemotherapy has been studied in women with operable breast cancer by Fisher [38, 39] and Wolmark [40]. Although there was no improvement in survival when AC chemotherapy was given preoperatively [40], there was an improvement in tumor response and in the use of breast conservation. Neoadjuvant chemotherapy in patients with breast cancer results in high response rates and has been used with the purpose of reducing tumor volume and achieving breast conservation in individuals who initially require mastectomy.

The clinical rationale for the use of thermotherapy and chemotherapy (thermochemotherapy) has been established in prior studies as described by Hornback [41], Dahl and Mella [42], Issels [43], Falk and Issels [44], Civadalli [45], Urano [46], and Englehardt [47]. Thermotherapy can potentially increase the cytotoxicity of certain chemotherapy agents by inhibition of cell repair mechanisms and by increasing the permeability of the cell membrane to allow more drug to destroy the tumor cells [41, p. 65]. Maximum effectiveness of heat and drug can be achieved when the heat is given within a few hours after chemotherapy is administered, which is based on the average distribution time of chemotherapy in humans being about 4 hours [41, p. 100].

In the study described in this chapter, it is hypothesized that for patients with breast carcinoma, the combination of focused microwave thermotherapy and neoadjuvant chemotherapy could improve tumor response rates in terms of tumor shrinkage and may contribute to increased breast conservation rates in a population judged to require mastectomy at the time of initial clinical presentation.

9.2 PATIENTS AND METHODS

Between November 2002 and May 2004, patients with primary invasive T2 (>2- to 5-cm clinical diameter), or T3 (> 5-cm clinical diameter) breast carcinomas seen at (1) University of Oklahoma, Oklahoma City; (2) Harbor-UCLA Medical Center, Torrance, California; (3) Comprehensive Breast Center, Coral Springs, Florida; (4) Mroz-Baier Breast Care Center, Memphis, Tennessee; (5) Pearl Place, Tacoma, Washington; (6) St. Joseph's Hospital, Orange, California; (7) Breast Care Specialists, Norfolk, Virginia; (8) Breast Care, Las Vegas, Nevada; and (9) Carolina Surgery, Gastonia, North Carolina were invited to participate in this FDA-approved prospective randomized open clinical study of the use of focused microwave thermotherapy in combination with neoadjuvant anthracycline-based chemotherapy (thermochemotherapy) compared with neoadjuvant anthracycline-based chemotherapy alone, as depicted in Figure 9.1. This study was approved and monitored by the Human Subjects Committee at each participating institution. Eligibility criteria included: (1) Karnofski performance status > 70%, (2) core needle biopsy-proven invasive breast cancer, (3) patient was a candidate for mastectomy and was eligible for neoadjuvant chemotherapy treatment, (4) visible tumor measurable by clinical exam and by ultrasound, (5) absence of involvement of the skin or pectoralis muscle. All patients were required to undergo counseling and sign written, informed consent.

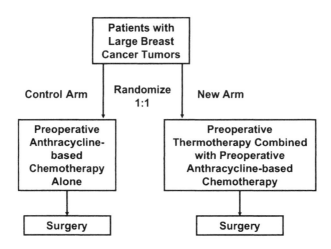

Figure 9.1 Clinical protocol for randomized study comparing preoperative thermochemotherapy with preoperative chemotherapy alone for patients with large breast cancer tumors [29].

Specific exclusion criteria were pregnancy, breast-feeding, and presence of breast implants, pacemakers, or defibrillators. Other exclusion criteria were known bleeding diathesis, laboratory evidence of coagulopathy (prothrombin time, international normalized ratio (PT, INR) > 1.5; partial thromboplastin time (PTT) > 1.5), thrombocytopenia (platelet count < 100,000/mm^3), anticoagulant therapy, or evidence of chronic liver disease or renal failure.

Patients randomized to thermochemotherapy were scheduled to receive four cycles (every 21 days) of AC chemotherapy (doxorubicin at 60 mg/m^2 and cyclophosphamide at 600 mg/m^2) and two treatments of focused microwave thermotherapy concomitantly with the first two cycles of AC chemotherapy – thermotherapy was to be administered starting within 1 to 4 hours, but no later than 36 hours, after chemotherapy was infused. Microwave treatment was performed on an outpatient basis using local anesthesia with patients in the prone position. A two-channel 915-MHz focused microwave adaptive phased array thermotherapy system (Microfocus APA-1000 Breast Thermotherapy System (refer to Figure 1.15), Celsion (Canada) Limited) was used in this study. This minimally invasive treatment system produces a wide focused microwave field in the compressed breast to heat and destroy high-water high-ion content breast carcinomas and tumor cells in the margins up to about 8 to 10 cm in maximum dimension as described in Chapter 8 [23]. Figure 9.2 shows a patient receiving wide-field focused microwave thermotherapy. Referring to Figure 8.2, a closed-

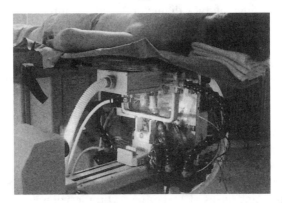

Figure 9.2 Photograph of a patient receiving adaptive phased array focused microwave thermotherapy. (Photograph courtesy of Celsion (Canada) Limited.)

end plastic catheter with a metal introducer was inserted into the primary tumor under ultrasound guidance, and a combination E-field focusing sensor and fiber-optic temperature sensor was inserted in the catheter to focus the microwaves and measure the tumor temperature during thermotherapy.

Seven thermocouple temperature sensors were taped to the skin and nipple to monitor the surface temperature during thermotherapy. During thermotherapy treatment, the amount of breast compression, focused microwave power, and air-cooling of the skin were adjusted to decrease thermotherapy-related side effects. The maximum allowed thermotherapy treatment time was 60 minutes in each of two treatments in this study. The clinical rationale, technology, and technique used for this approach of focused microwave phased array (FMPA) thermotherapy for tumor ablation was previously described in Chapter 8 [20, 22, 23, 29, 30].

Thermal dose is accumulated in the tumor at a faster rate during tumor temperature increases above 43°C, which is referred to as cumulative equivalent minutes thermal dose CEM(43°C) [35]. Using (1.1), the CEM(43°C) was calculated from the measured temperatures recorded by the sensor in the tumor – the desired tumor thermal dose during active microwave heating for this study was in the range of 80 to 120 CEM(43°C) in each of two treatments at tumor temperatures in the range of 44° to 46°C. Based on two such thermotherapy treatments, the desired cumulative thermal dose would range between 160 to 240 CEM(43°C) and should provide a significant tumor response, because a heat-alone cumulative thermal dose of 210 minutes or greater (relative to 43°C) is predictive of 100% necrosis for invasive breast carcinomas [22].

Breast tumor response was quantified by clinical exam in two dimensions (area = product of two diameters) and by ultrasound measurements in three dimensions (elliptical volume = length × width × depth × 0.524). Tumor cell kill was based on necrosis determined by hematoxylin and eosin (H&E) histological sections from the excised breast tumor. Necrosis was estimated and expressed as a percentage of necrotic tumor areas in relation to necrotic and viable tumor areas.

Statistical differences between thermochemotherapy and chemotherapy alone groups were quantified using a Student t test, a Mann-Whitney test, and Fisher's exact test (InStat, GraphPad Software, Inc.), as appropriate. All tests were two-sided and P values ≤ 0.05 were considered statistically significant. For this small study, the median value tends to be less sensitive to outlier values and is used to compare tumor volume reduction for the two groups. Comparisons were made on study subjects who completed the assigned treatment.

9.3 RESULTS

A total of 34 adult female patients with T2 and T3 invasive breast cancer were enrolled in this study [29]. Seventeen (17) patients were randomized to thermochemotherapy and 17 were randomized to chemotherapy-alone. A larger enrollment was planned, but due to changes in standard-of-care neoadjuvant chemotherapy that occurred at some of the participating institutions during this period of time, the trial was closed early; however, key findings are discussed below.

Analyzable data are available for 15 of 17 patients in the thermochemotherapy arm, and 13 of 17 subjects in the chemotherapy alone arm. Six patients were withdrawn from the study because their chemotherapy regimen was altered to a nonanthracycline-based chemotherapy regimen after enrollment, leaving a total of 28 study subjects for analysis.

Patient clinical and tumor characteristics at the time of enrollment in the study for the two groups are depicted and compared in Table 9.2. At enrollment, based on clinical examination the median value of tumor size in the thermochemotherapy arm and in the chemotherapy alone arm were 5.13 cm and 3.5 cm, respectively, ($P = 0.05$, Mann-Whitney two-sided test); based on ultrasound measurements, the median value of tumor volume in the thermochemotherapy arm and the chemotherapy-alone arm were 10.31 cc and 4.13 cc respectively, ($P = 0.02$, Mann-Whitney two-sided test). Thus, prior to treatment, tumors were significantly larger in the thermochemotherapy arm compared to the chemotherapy arm.

The measured temperatures in the tumor and on the skin surface during a typical focused microwave thermotherapy treatment are shown in Figure 9.3. The corresponding microwave power versus time is shown in Figure 9.4. In the thermochemotherapy arm, the mean peak tumor temperature achieved was 45.0°C (range 44.6° to 46.5°C) and the mean tumor thermal dose was 150.6 CEM(43°C) (range 0 to 233.9 CEM(43°C)). Mean thermotherapy treatment time was 34.8 minutes per each treatment. Tumor response in terms of tumor volume reduction following thermochemotherapy and chemotherapy alone is summarized in Table 9.3. After thermochemotherapy, but prior to surgery, the mean tumor diameter was 1.59 cm (range 0 to 3.5 cm) and mean tumor volume was 3.13 cc (range 0 to 16.43 cc) based on ultrasound – mean tumor volume reduction compared to the volume at enrollment was 69.6% with median value 88.4% ($n = 14$). After chemotherapy alone but prior to surgery, the mean tumor diameter based on ultrasound was 1.97 cm (range 1.0 to 4.54 cm) and mean tumor volume was 2.3 cc (range 0.24 to 7.65 cc) based on ultrasound – mean tumor reduction was 50.5% with median value 58.8% ($n = 10$) based on ultrasound-measured volume. Figure 9.5 provides a comparison of the tumor volume reduction in both arms based on ultrasound measurements – the data are ordered from the greatest tumor reduction to least tumor reduction with negative values indicating tumor growth. The median value of absolute tumor volume reduction for

Table 9.2
Demographics and Tumor Characteristics of the Overall Study Population [29]

	Thermochemotherapy	Chemotherapy Alone	P-value
N	15	13	
Age, years			
Mean	45.1	45.8	0.86
Range	26-72	32-79	
Menopausal status			1.0
Pre	11	9	
Post	4	4	
Clinical tumor size at enrollment			
Mean, cm	5.26	4.19	0.17
Median, cm	5.13	3.5	0.05
Range, cm	2.4-7.5	2.5-9.0	
Clinical tumor classification at enrollment			0.15
T2	6	9	
T3	9	4	
Clinical nodal status at enrollment			0.7
Negative	10	7	
Positive	5	6	
Tumor histology			0.33
IDC	11	12	
ILC	3	0	
IMDLC	1	1	
Ultrasound measured tumor size at enrollment			
Mean, cm	3.65	2.69	0.10
Median, cm	3.24	2.39	0.07
Range, cm	2.0-7.8	1.22-6.5	
Ultrasound measured tumor volume at enrollment			
Mean, cm	22.098	10.17	0.18
Median, cm	10.31	4.13	0.02
Range, cm	1.85-90.98	0.39-78.68	

IDC=invasive ductal carcinoma, ILC=invasive lobular carcinoma, IMDLC=invasive mixed.

preoperative thermochemotherapy was significant relative to preoperative chemotherapy alone (88.4% versus 58.8% respectively, $P = 0.048$, Mann-Whitney two-sided test). As a result of the thermochemotherapy treatments,

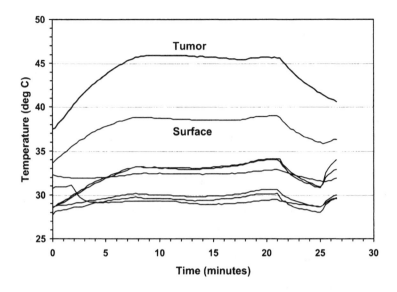

Figure 9.3 Measured tumor and surface temperatures during an adaptive phased array focused microwave treatment (case 1406-t1).

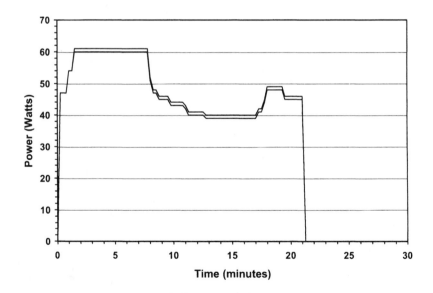

Figure 9.4 Microwave power for the two channels during an adaptive phased array focused microwave treatment (case 1406-t1).

14 of 15 (93.3%) patients had sufficient tumor volume reduction in the breast to become eligible for breast conservation based on cosmetic considerations by the judgment of the surgeon, and 11 of 15 (73.3%) patients actually received breast conservation surgery. In the chemotherapy-alone arm, 12 of 13 (92.3%) of patients were eligible and actually received breast conservation surgery ($P = 0.33$ by Fisher's exact test). Mean pathologic tumor necrosis by volume was 70.5% (range 0% to 100%, $n = 14$) with two patients having a complete pathologic response among those receiving thermochemotherapy,

Table 9.3

Percent Tumor Shrinkage Based on Ultrasound Measurements Prior to Surgery for Chemotherapy-Alone Arm and Thermochemotherapy Arm of the Overall Study Population [29]

Study Parameter	Thermochemotherapy $n = 14$	Chemotherapy Alone, $n = 10$	P-value
Tumor shrinkage based on ultrasound volume			
Mean	69.6%	50.5%	0.358
Median	88.4%	58.8%	0.048
Range	-99.9 to 100%	-66.7 to 99.1%	

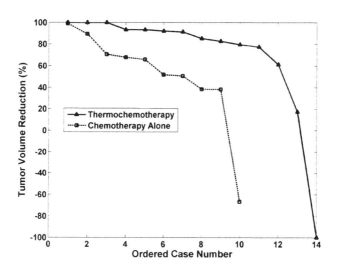

Figure 9.5 Comparison of ultrasound-measured tumor volume reduction based on three-dimensional measurements for patients treated in the two arms of the study [29]. The cases have been ordered by decreasing tumor response.

Table 9.4
Side Effects for the Thermochemotherapy Arm of the Overall Study Population [29]

Side Effect	Number of Occurences
Erythema	6 of 26 (25.1%) treatments
Skin burn	5 of 26 (19.2%) treatments
Level of discomfort during thermotherapy None	5 of 22 (22.7%) treatments
Mild	11 of 22 (50.0%) treatments
Moderate	5 of 22 (22.7%) treatments
Intolerable	1 of 22 (4.5%) treatments

and in the chemotherapy-alone arm mean pathologic tumor necrosis by volume was 45.7% (range 0% to 100%, $n = 7$), $P = 0.41$. Side effects caused by thermotherapy are summarized in Table 9.4. Erythema (temporary skin redness) occurred in 6 of 26 (23.1%) treatments. A skin burn (less than 1.5 cm diameter) in the microwave treatment field occurred in 5 of 26 (19.2%) treatments - in each case the peak skin temperature exceeded 40°C (range 40.4° to 42.5°C). The subjects' level of discomfort with thermotherapy reported for 22 treatments was 5 of 22 (22.7%) no discomfort, 11 of 22 (50.0%) mild discomfort, 5 of 22 (22.7%) moderate discomfort, and 1 of 22 (4.5%) intolerable discomfort.

Information on thermochemotherapy-treated patients ($n = 7$) with clinical tumors 3.5 cm or larger is now summarized. At enrollment, based on clinical examination the mean largest perpendicular tumor diameter was 5.68 cm (range 3.5 to 7.5 cm) and the mean tumor area was 27.68 cm^2 (range 16 to 45 cm^2). Based on ultrasound measurements at enrollment, mean tumor diameter was 2.88 cm (range 2.0 to 4.4 cm) and mean tumor volume was 12.40 cc (range 2.48 to 31.08 cc). All patients completed four cycles of chemotherapy prior to surgery with the exception of case 1410 in which the tumor was judged (based on ultrasound measurements) to be growing after the first three chemotherapy treatments were administered, and the patient then received a mastectomy (pathology of the mastectomy specimen revealed 85% necrosis in the tumor). These patients received two thermotherapy treatments each, but 2 of 14 (14.3%) treatments were stopped prior to completing a full thermal dose because of patient discomfort from the thermotherapy treatment. The mean peak tumor temperature achieved for the seven patients was 45.5°C (range 44.6° to 46.5°C) and the mean tumor thermal dose was 190.4 minutes (range 108.5 to 233.9 minutes). For the combination of two thermotherapy treatments, the desired minimum CEM(43°C) dose of 160 minutes or greater was achieved in 5 of 7 (71.4%) patients.

Based on clinical examination, preoperative tumor size measurements after thermochemotherapy were a mean tumor diameter of 2.50 cm (range 0 to 4.0 cm) and a mean tumor area of 6.50 cm^2 (range 0 to 16 cm^2). Based on ultrasound measurements, the mean tumor diameter was 1.49 cm (range 0 to 1.87 cm) and the mean tumor volume was 1.82 cc (range 0 to 4.97 cc). Comparing tumor dimensions at enrollment with those prior to surgery, the mean tumor size reduction based on clinical tumor area was 73.9% (range 20% to 100%), and the mean tumor size reduction based on ultrasound tumor volume was 60.5% (range −99.9% to 100%). Following thermochemotherapy treatments, 6 of 7 (85.7%) patients became eligible for breast conservation based on expected cosmesis judged by the surgeon. One of these subjects (case 1403) had a significant response from thermochemotherapy and was offered the option for breast conservation, but the patient chose to have a mastectomy – based on H&E pathology from the mastectomy specimen there was 99% tumor necrosis by volume. Five of 7 (71.4%) patients actually received breast conservation and 2 patients received mastectomy. For the 7 patients, mean pathologic tumor necrosis by volume was 81.3% (range 20% to 100%). One of 7 (14.3%) patients had a complete clinical tumor response and this patient also had a complete pathologic response.

9.4 DISCUSSION

In this clinical study of patients with T2 and T3 tumors judged to require mastectomy as the local surgical treatment for breast cancer, preoperative focused microwave phased array thermotherapy in combination with preoperative AC chemotherapy appears to be effective and safe. The combined use of thermotherapy and AC chemotherapy in the overall group provided a greater tumor response in terms of tumor volume reduction — median tumor size reduction based on ultrasound-measured volume was 88.4% in the thermochemotherapy arm compared to 58.8% in the chemotherapy-alone control arm, and was statistically significant ($P = 0.048$).

In the subset group with tumors 3.5 cm or greater, the target thermal dose of 160 equivalent minutes was achieved in 5 of 7 (71.4%) patients, and the desired peak tumor temperature in the range of 44° to 46°C was achieved in 7 of 7 (100%) patients. Sufficient tumor volume reduction in the breast was achieved such that 6 of 7 (85.7%) patients became eligible for breast conservation. The actual rate of conversion to breast conservation using combined thermotherapy with neoadjuvant AC chemotherapy was 5 of 7 (71.4%) patients. Compared to the literature, patients with large invasive breast carcinoma receiving neoadjuvant chemotherapy alone, conversion rates to breast conservation vary from about 20% to 60% depending on tumor size at presentation as described by Fisher [38], Schwartz [48], and Kaufmann [49]

This small study serves as proof of the principle for this neoadjuvant combined approach of focused microwave thermotherapy and chemotherapy. As of this date, further clinical testing of neoadjuvant focused microwave thermochemotherapy [50] would be necessary to demonstrate an objective clinical response (clinically measurable improved patient outcome) such as increased tumor response, increased use of breast conservation, or decreased recurrence rates.

As discussed in Chapter 8, mastectomy rates have risen significantly in the last 3 years based on a study published by Katipamula [51] — at the Mayo Clinic in Rochester, MN, the rate of patients receiving mastectomy was 43% in 2006 compared to 30% in 2003. Treatments, such as neoadjuvant thermochemotherapy as described in this chapter, that could reduce the rate of mastectomy and increase the use of breast conservation are desirable.

Ablation techniques such as radiofrequency, interstitial laser photocoagulation, focused ultrasound, and cryotherapy have been used to treat invasive breast carcinomas up to about 3 cm in maximum dimension and would have limited application for patients with breast carcinomas larger than 3 cm, whereas focused microwaves with wide-field applicators can treat breast carcinomas up to about 8 to 10 cm in maximum dimension [20, 23, 29, 30].

For a subset group of patients with large tumors 3.5 cm or greater, 7 of 7 (100%) patients receiving thermochemotherapy exhibited various degrees of tumor necrosis that ranged from 20% to 100%. Based on H&E staining, complete (100%) necrosis of breast cancer was achieved in 1 of 7 (14.3%) patients with large tumors. An increased thermal dose than that used in this study would be required to achieve complete pathological tumor response. Tumor cell kill measured by other pathologic testing such as nicotinamide adenine dinucleotide-diaphorase (NADH-d) and immunohistochemistry, as used in other breast tumor ablation studies [5, 20], were not evaluated in this study. It is hypothesized that preoperative focused microwave thermotherapy used in combination with preoperative AC chemotherapy might provide an enhanced therapeutic effect on large carcinomas in the breast and might provide a means for improving the rate of breast conservation by increasing tumor shrinkage and increasing pathologic tumor cell kill. This will be the subject of a future study.

Potential weaknesses in the current study design may be responsible for a lack of clinical efficacy and may be related to the absence of very well defined or strict criteria for the selection of breast conservation surgery versus mastectomy, as this was based solely on experienced surgeons' judgments. Another possible explanation is that tumor size was such that it required only reduction of a small volume to permit breast conservation surgery. Furthermore, there is an apparent difference in pretreatment tumor size between the experimental and control arms, tumors being larger in the experimental, thermotherapy arm.

There is no significantly increased morbidity with the addition of thermotherapy to the standard systemic chemotherapy. The treatment was well tolerated in the vast majority of patients. Self-limited erythema occurred commonly after thermotherapy and only in a small number of patients was this side effect followed by clinical evidence of a burn area in the skin. Skin burn in most patients was treated with topical treatments, but required excision of the small burn area in two cases. Close attention to the technique of delivery of thermotherapy with particular interest in the thermal dose delivered to the skin and to patient symptomatology during therapy are essential to avoid significant hyperthermia-related side effects. Improvements in the cooling system may also facilitate the delivery of tumoricidal thermal doses while keeping normal tissues safe.

The risk of local recurrence is an important factor that warrants consideration. A shortcoming of this study is the limited followup of the patients in this series. Larger studies and long-term followup is required to accurately estimate the incidence of local recurrence after breast conservation surgery following neoadjuvant thermochemotherapy.

Based on this study, a future clinical trial for patients with large breast carcinomas, approximately 3.5 cm or greater, could explore the use of preoperative thermotherapy in combination with preoperative anthracycline-based (doxorubicin or epirubicin) combination chemotherapy versus preoperative anthracycline-based combination chemotherapy alone [51]. This future study could examine the role of thermochemotherapy for increased tumor response and increased use of breast conservation.

In conclusion, in this small randomized study, neoadjuvant thermotherapy in combination with chemotherapy is more effective than neoadjuvant chemotherapy in reducing tumor burden. Both neoadjuvant methods similarly allow for breast conservation surgery in selected women with invasive breast cancer who would have required total mastectomy at the time of initial presentation.

9.5 SUMMARY

Microwave energy is a promising method for breast tumor ablation and thermotherapy is synergistic with chemotherapy for treatment of carcinomas. The aim of the study described in this chapter was to investigate whether preoperative adaptive phased array focused microwave thermotherapy, in combination with neoadjuvant anthracycline-based chemotherapy, could be safely administered and provide an increased local tumor response for patients with breast carcinomas. Patients with invasive breast carcinoma (T2 and T3) that were candidates for mastectomy and were not candidates for breast conservation therapy at the time of presentation were invited in the study. Tumor dose was measured as cumulative thermal equivalent minutes (CEM(43°C)) of treatment relative to 43°C, with a goal of 160 to 240 CEM(43°C) total over the first two of four planned neoadjuvant AC (doxorubicin, cyclophosphamide) chemotherapy treatments. Outcomes measured in this study were tumor volume reduction, breast conservation rate, pathologic tumor necrosis, and thermotherapy-related side effects. Fifteen patients received focused microwave thermotherapy and anthracycline-based chemotherapy prior to surgery, and 13 patients received anthracycline-based chemotherapy alone prior to surgery. In the thermochemotherapy arm, the mean peak tumor temperature achieved was 45.0°C and the mean tumor CEM(43°C) was 150.6 minutes. Tumor volume reduction based on the ultrasound-measured median tumor volume was 88.4% in the thermochemotherapy arm compared to 58.8% in the chemotherapy-alone control arm and was statistically significant ($P = 0.048$). Preoperative focused microwave thermotherapy in combination with neoadjuvant AC chemotherapy can be administered safely and improves tumor response compared to AC chemotherapy alone for patients with T2 and T3 breast carcinomas with minimal morbidity.

References

[1] Hall-Craggs, M.A., "Interventional MRI of the Breast: Minimally Invasive Therapy," *Eur Radiol*, Vol. 10, No. 1, 2000, pp. 59-62.

[2] Singletary, S.E., "Minimally Invasive Techniques in Breast Cancer Treatment," *Seminars in Surgical Oncology,* Vol. 20, 2001, 246-250.

[3] Noguchi, M., "Minimally Invasive Surgery for Small Breast Cancer," *J Surg Oncol,* Vol. 84, 2003, pp. 94-101.

[4] Agnese, D.M., and W.E. Burak, "Ablative Approaches to the Minimally Invasive Treatment of Breast Cancer," *Cancer Journal*, Vol. 11, 2005, pp. 77-82.

[5] Huston, T.L, and R.M. Simmons, "Ablative Therapies for the Treatment of Malignant Diseases of the Breast," *Am J Surg*, Vol. 189, 2005, pp. 694-701.

[6] van Esser, S., M.A.A.J. van den Bosch, P.J. van Diest, W.Th.M. Mali, I.H.M. Borel Rinkes, R. van Hillegersberg, "Minimally Invasive Ablative Therapies for Invasive Breast Carcinomas: An Overview of Current Literature," *World J Surgery*, 2007, Published online October 24, 2007.

[7] Jeffrey, S.S., R.L. Birdwell, D.M. Ikeda, et al., "Radiofrequency Ablation of Breast Cancer: First Report of an Emerging Technology," *Arch Surg,* Vol. 134, No. 10, 1999, pp. 1064-1068.

[8] Izzo, F., et al., "Radiofrequency Ablation in Patients with Primary Breast Carcinoma. A Pilot Study of 26 Patients," *Cancer*, Vol. 92, No. 8, 2001, pp. 2036-2044.

[9] Singletary, S.E., B.D. Fornage, N. Sneige, et al., "Radiofrequency Ablation of Early-Stage Invasive Breast Tumors: an Overview," *Cancer J*, Vol. 8, 2002, pp. 177-180.

[10] Burak, W.E., D.M. Agnese, S.P. Povoski, et al., "Radiofrequency Ablation of Invasive Breast Carcinoma Followed by Delayed Surgical Excision," *Cancer*, Vol. 98, 2003, pp. 1369-1376.

[11] Fornage, B.D., et al., "Small ($\leq$2-cm) Breast Cancer Treated with US-Guided Radiofrequency Ablation: Feasibility Study," *Radiology*, Vol. 231, No. 1, 2004, pp. 215-226.

[12] Ross, M.I., and B.D. Fornage,"Radiofrequency Ablation of Early-Stage Breast Cancer," In: Chapter 9 of *Radiofrequency Ablation for Cancer: Current Indications, Techniques, and Outcomes,* Ellis, L.M., S.A. Curley, and K.K. Tanabe, (eds.), New York: Springer-Verlag, 2004.

[13] Dowlatshashi, K., D.S. Francescatti, K.J. Bloom, "Laser Therapy for Small Breast Cancers," *Am J Surgery*, Vol. 184, 2002, pp. 359-363.

[14] Dowlatshashi, K., J.J. Dieschbourg, K.J. Bloom, "Laser Therapy of Breast Cancer with 3-Year Follow-Up," *The Breast Journal,* Vol. 10, No. 3, 2004, pp. 240-243.

[15] Huber, P.E., et al., "A New Noninvasive Approach in Breast Cancer Therapy Using Magnetic Resonance Imaging-Guided Focused Ultrasound Surgery," *Cancer Res*, Vol. 61, 2001, pp. 8441-8447.

[16] Wu, F., "Extracorporeal High Intensity Focused Ultrasound Treatment for Patients with Breast Cancer," *Breast Cancer Res Treat*, Vol. 92, 2005, pp. 51-60.

[17] Hynynen, K., "Focused Ultrasound Surgery Guided by MRI," *Science & Medicine*, Vol. 3, No. 5, 1996, pp. 62-71.

[18] Pfleiderer, S.O., M.G. Freesmeyer, C. Marx, R. Kuhne-Heid, A. Schneider, and W.A. Kaiser, "Cryotherapy of Breast Cancer Under Ultrasound Guidance: Initial Results and Limitations," *Eur Radiol,* Vol. 12, 2002, pp. 3009-3014.

[19] Sabel, M.S., C.S. Kaufman, P. Whitworth, H. Chang, L.H. Stocks, R. Simmons, and M. Schultz, "Cryoablation of Early-Stage Breast Cancer: Work-in-Progress Report of a Multi-Institutional Trial," *Ann Surg Oncol*, Vol. 11, No. 5, 2004, pp. 542-549.

[20] Gardner, R.A., H.I. Vargas, J.B. Block, C.L. Vogel, A.J. Fenn, G.V. Kuehl, and M. Doval, "Focused Microwave Phased Array Thermotherapy for Primary Breast Cancer," *Ann Surg Oncol*, Vol. 9, No. 4, 2002, pp. 326-332.

[21] Vargas, H.I., W.C. Dooley, R.A. Gardner, K.D. Gonzalez, S.H. Heywang-Kobrunner, and A.J. Fenn, "Success of Sentinel Lymph Node Mapping After Breast Cancer Ablation With Focused Microwave Phased Array Thermotherapy," *Am J Surg*, Vol. 186, 2003, pp. 330-332.

[22] Vargas, H.I., W.C. Dooley, R.A. Gardner, K.D. Gonzalez, S.H. Heywang-Kobrunner, and A.J. Fenn, "Focused Microwave Phased Array Thermotherapy for Ablation of Early-Stage Breast Cancer: Results of Thermal Dose Escalation," *Ann Surg Oncol*, Vol. 11, No.

2, 2004, pp. 139-146.

[23] Fenn, A.J., *Breast Cancer Treatment by Focused Microwave Thermotherapy*, Sudbury, MA: Jones and Bartlett, 2007.

[24] Fenn, A.J., G.L. Wolf, and R.M. Fogle, "An Adaptive Phased Array for Targeted Heating of Deep Tumors in Intact Breast: Animal Study Results," *Int J Hyperthermia*, Vol. 15, No. 1, 1999, pp. 45-61.

[25] Gavrilov, L.R., J.W. Hand, J.W. Hopewell, and A.J. Fenn, "Pre-clinical Evaluation of a Two-Channel Microwave Hyperthermia System with Adaptive Phase Control in a Large Animal," *Int J Hyperthermia*, Vol. 15, No. 6, 1999, pp. 495-507.

[26] Fenn, A.J., "Adaptive Hyperthermia for Improved Thermal Dose Distribution," In: *Radiation Research: A Twentieth Century Perspective*, Vol. 1 (Congress Abstracts), Chapman J.D., W.C. Dewey, G.F. Whitmore, (eds.), San Diego, Calif.: Academic Press, 1991, p. 290.

[27] Fenn, A.J., A.Y. Cheung, and H. Cao, "Adaptive Focusing Experiments for Minimally Invasive Monopole Phased Arrays in Hyperthermia Treatment of Breast Cancer," *16th Annual IEEE Engineering in Medicine and Biology Society Int Conf*, Baltimore, Maryland: November 3-6, 1994, pp. 766-767.

[28] Fenn, A.J., "Minimally Invasive Monopole Phased Arrays for Hyperthermia Treatment of Breast Cancer," In: *Proc. 1994 Int. Symp. on Antennas*, Nice, France: November 8-10, 1994, pp. 418-421.

[29] Vargas, H.I., W.C. Dooley, A.J. Fenn, M.B. Tomaselli, and J.K. Harness, "Study of Preoperative Focused Microwave Phased Array Thermotherapy in Combination with Neoadjuvant Anthracycline-Based Chemotherapy for Large Breast Carcinomas," *Cancer Therapy*, Vol. 5, 2007, published online (www.cancer-therapy.org), November 25, 2007, pp. 401-408.

[30] Dooley, W.C., H.I. Vargas, A.J. Fenn, M.B. Tomaselli, and J.K. Harness, "Randomized Study of Preoperative Focused Microwave Phased Array Thermotherapy for Early-Stage Invasive Breast Cancer," *Cancer Therapy*, Vol. 6, 2008, published online (www.cancer-therapy.org), August 25, 2008, pp. 395-408.

[31] Campbell, A.M., and D.V. Land, "Dielectric Properties of Female Human Breast Tissue Measured in Vitro at 3.2 GHz," *Phys Med Biol*, Vol. 37, No. 1, 1992, pp. 193-210.

[32] Joines, W.T., Y. Zhang, C. Li, and R.L. Jirtle, "The Measured Electrical Properties of Normal and Malignant Human Tissues From 50 to 900 MHz," *Med Phys*, Vol. 21, No. 4, 1994, pp. 547-550.

[33] Burdette, E.C. "Electromagnetic and Acoustical Properties of Tissue," In: *Physical Aspects of Hyperthermia*, Nussbaum, G.H., (ed.), AAPM Medical Physics Monographs, No. 8, 1982, pp. 105-130.

[34] Lazebnik, M., et al., "A Large-Scale Study of the Ultrawideband Microwave Dielectric Properties of Normal, Benign, and Malignant Breast Tissues Obtained from Cancer Surgeries," *Phys Med Biol*, Vol. 52, 2007, pp. 6093-6115.

[35] Sapareto, S.A., and W.C. Dewey, "Thermal Dose Determination in Cancer Therapy," *Int J Rad Oncol Biol Phys*, Vol. 10, 1984, pp. 787-800.

[36] Hamilton, A., and G. Hortobagyi, "Chemotherapy: What Progress in the Last 5 Years?" *J Clin Oncol*, Vol. 23, 2005, pp. 1760-1775.

[37] Kaufmann, M., et al., "Recommendations from an International Expert Panel on the Use of Neoadjuvant (Primary) Systemic Treatment of Operable Breast Cancer, an Update," *J Clin Oncol*, Vol. 24, No. 12, 2006, 1940-1949.

[38] Fisher, B., et al., "Effect of Preoperative Chemotherapy on Local-Regional Disease in Women with Operable Breast Cancer: Findings from the National Surgical Adjuvant Breast and Bowel Project B-18," *J Clin Oncol*, Vol. 15, No. 7, 1997, pp. 2483-2493.

[39] Fisher, B., et al., "Effect of Preoperative Chemotherapy in the Outcome of Women with Operable Breast Cancer," *J Clin Oncol*, Vol. 16, No. 8, 1998, pp. 2672-2685.

[40] Wolmark, N., J. Wang, E. Mamounas, J. Bryant, and B. Fisher, "Preoperative Chemotherapy in Patients with Operable Breast Cancer: Nine-Year Results from National Surgical Adjuvant Breast and Bowel Project B-18," *J Natl Cancer Inst Monographs,* No. 30, 2001, pp. 96-102.

[41] Hornback, N., *Hyperthermia and Cancer: Volume I,* Boca Raton, Florida: CRC Press, 1984, pp. 65-75, 94-104.

[42] Dahl, O., and O. Mella, "Hyperthermia and Chemotherapeutic Agents," In: Chapter 5 of *An Introduction to the Practical Aspects of Clinical Hyperthermia,* Field, S.B. and J.W. Hand, (eds.), New York: Taylor & Francis, 1990, pp. 108-142.

[43] Issels, R., "Clinical Rationale for Thermochemotherapy," In: Chapter 2 of *Thermoradio-therapy and Thermochemotherapy, Vol. 2, Clinical Applications,*, Seegenschmiedt, M.H., P. Fessenden, and C.C. Vernon, (eds.), Berlin: Springer, 1995, pp. 25-33.

[44] Falk, M.H., and R.D. Issels, "Hyperthermia in Oncology," *Int J Hyperthermia,* Vol. 17, No. 1, 2001, pp. 1-18.

[45] Cividalli, A., et al., "Hyperthermia and Paclitaxel - Epirubicin Chemotherapy: Enhanced Cytotoxic Effect in a Murine Mammary Adenocarcinoma," *Int J Hyperthermia,* Vol. 16, No. 1, 2000, pp. 61-71.

[46] Urano, M., J. Begley, and R. Reynolds, "Interaction Between Adriamycin and Hyperthermia: Growth-Phase-Dependent Thermal Sensitization," *Int J Hyperthermia,* Vol. 10, No. 6, 1994, pp. 817-826.

[47] Englehardt, R., "Hyperthermia and Drugs," In: *Hyperthermia and the Therapy of Malignant Tumors,* Streffer, C., (ed.), Berlin: Springer, 1987, pp. 136-176.

[48] Schwartz, G.F., A.J. Meltzer, and E.A. Lucarelli, "Breast Conservation After Neoadjuvant Chemotherapy for Stage II Carcinoma of the Breast," *J Am Coll Surg*, Vol. 201, 2005, pp. 327-334.

[49] Kaufmann, P., C.E. Dauphine, M.P. Vargas, M.L. Burla, N.M. Isaac, K.D. Gonzalez, D. Rosing, and H.I. Vargas, "Success of Neoadjuvant Chemotherapy in Conversion of Mastectomy to Breast Conservation Surgery," *Am Surgeon,* Vol. 72, No. 10, 2006, pp. 935-938.

[50] Vargas, H.I., W.C. Dooley, A.J. Fenn, M.B. Tomaselli, and J.K. Harness, "A Protocol for

Focused Microwave Thermotherapy in Combination with Neoadjuvant Anthracycline-Based Chemotherapy for Investigating Tumor Response of Large Invasive Breast Carcinomas," *American J Clin Oncol*, Abstract, October, 2008.

[51] Katipamula, R., et al., "Trends in Mastectomy Rates at the Mayo Clinic Rochester: Effects of Surgical Year and Preoperative MRI," 2008 American Society of Clinical Oncology 44th Annual Meeting, May 30-June 3, 2008, Chicago, IL, *Journal of Clinical Oncology*, Supplement, Vol. 26, No. 15S, Part I of II, 2008, Abstract 509, p. 9s.

10

Future Studies of Adaptive Phased Arrays for Cancer

In this book, an approach for treating cancer with minimally invasive adaptive electromagnetic phased arrays has been described. It is well known that heat can kill cancer cells when used alone or in combination with radiation therapy or chemotherapy, but it can be difficult to deliver the necessary heat to the tumor without burning the skin and surrounding healthy tissues. It has been shown in this book with measurements and simulations that an adaptive phased array can potentially provide reliable heating of deep tumors in the torso while nulling the surface fields [1-11]. It has been shown with preclinical and clinical studies that an adaptive phased array focused microwave thermotherapy system can safely heat small to large breast cancer tumors at depth in the intact breast [12-24].

Sample matrix inversion and gradient-search algorithms for controlling the amplitude and phase of adaptive transmit phased array thermotherapy systems were reviewed in Chapter 2. Chapter 3 reviewed electromagnetic field theory and computational techniques for predicting the electric fields and specific absorption rate in homogeneous and heterogeneous tissues. Computation of the electromagnetic field can be accomplished in an approximate manner by ray tracing, by the method of moments, or more accurately by the finite-difference time-domain technique for solving Maxwell's equations. The bioheat equation and the finite-difference time-domain technique can also be used to compute the temperature distribution in heterogeneous tissues as discussed in Chapter 4. In Chapter 5, adaptive nulling was investigated as a means to reduce the electric fields at the surface of a phantom target. It was shown in a homogeneous lossy tissue medium that a radiation beam can be focused and maintained on a central tissue site, while the surface fields are nulled or minimized. Thermal modeling showed that tissue hot spots could be reduced by nulling the surface electric fields at radiofrequencies. Due to the finite null width of a null formed on the surface of a phantom target, the electric field is also reduced in the subsurface tissue to reduce tissue hot spots away from the tumor.

In Chapter 6, measurements in three types of phantoms with a clinical dipole phased array hyperthermia system modified to perform adaptive nulling and focusing were described. The measurements showed that adaptive nulls could be formed on the phantom surface and that the electric field delivered to a target region could be controlled and increased in amplitude. In Chapter 7, a monopole phased array waveguide applicator was described for deep heating of tumors in the torso. This monopole array could be used in adaptive phased array animal studies followed by human clinicals for treating deep-seated cancers. Chapter 8 reviewed the design and preclinical testing of an adaptive phased array focused microwave thermotherapy system for treating small to large tumors in the intact breast. Chapter 9 described details of one of the clinical trials for preoperative adaptive phased array focused microwave thermotherapy. In this study, thermotherapy in combination with preoperative chemotherapy for treating large invasive cancer tumors in the intact breast demonstrated increased tumor volume reduction compared to standard preoperative chemotherapy treatment. Some of the future studies that could be conducted for adaptive phased array thermotherapy for cancer treatment are listed below.

As discussed in Chapter 8, most of the clinical development of adaptive phased arrays by this author for cancer treatment have been for treating cancer in the intact breast. Since there is a wide variation in the type or stage of breast cancer, different protocols are necessary to address clinical needs. These future protocol topics, described in detail by the author elsewhere [22, pp. 183-205], are listed below.

- Breast Cancer (randomized studies of focused microwave thermother-apy)
  - Early-stage: Preoperative thermotherapy (ablation) to improve surgical margins, reduce or replace surgery, reduce local recur-rence, and improve cosmesis [24].
  - Large breast cancer tumors: Preoperative thermochemotherapy to increase tumor shrinkage (tumor volume reduction) and increase breast conservation [23, 25].
  - Recurrent chest wall: Thermotherapy in combination with radiation therapy to improve tumor response.

- Ductal carcinoma in situ (DCIS): Breast conserving surgery followed by thermotherapy followed by radiation therapy and tamoxifen to reduce cancer recurrence rate.
- Breast cancer prevention: Thermotherapy in combination with antiestrogen drugs to damage estrogen receptors.
- Cancer stem cell ablation: Thermotherapy to destroy breast cancer stem cells.

All of the above protocols for breast cancer will require randomized trials to demonstrate patient efficacy. For treatment of deep cancer tumors including bladder, brain, colon, head and neck, kidney, liver, lung, ovary, pancreas, prostate, rectum, stomach, and uterine, adaptive phased array thermotherapy could be considered in combination with radiation therapy, chemotherapy, or gene therapy, and are the subject of future studies.

References

[1] Fenn, A.J., "Adaptive Nulling Hyperthermia Array," US Patent No. 5,251,645, October 12, 1993.

[2] Fenn, A.J., "Adaptive Focusing and Nulling Hyperthermia Annular and Monopole Phased Array Applicators," US Patent No. 5,441,532, August 15, 1995.

[3] Fenn, A.J., "Adaptive Hyperthermia for Improved Thermal Dose Distribution," In: *Radiation Research: A Twentieth Century Perspective,* Vol. 1 (Congress Abstracts), Chapman J.D., W.C. Dewey, G.F. Whitmore, (eds.), San Diego, Calif.: Academic Press, 1991, p. 290.

[4] Fenn, A.J., and G.A. King, "Adaptive Nulling in the Hyperthermia Treatment of Cancer," *The Lincoln Laboratory Journal,* Lincoln Laboratory, Massachusetts Institute of Technology, Vol. 5, No. 2, 1992, pp. 223-240.

[5] Fenn, A.J., C.J. Diederich, and P.R. Stauffer, "An Adaptive-Focusing Algorithm for a Microwave Planar Phased-Array Hyperthermia System," *The Lincoln Laboratory Journal,* Lincoln Laboratory, Massachusetts Institute of Technology, Vol. 6, No. 2, 1993, pp. 269-288.

[6] Fenn A.J., and G.A. King, "Experimental Investigation of an Adaptive Feedback Algorithm for Hot Spot Reduction in Radio-Frequency Phased-Array Hyperthermia," *IEEE Trans Biomed Eng.,* Vol. 43, No. 3, 1994, pp. 273-280.

[7] Fenn, A.J., and G.A. King, "Adaptive Radio Frequency Hyperthermia Phased Array System for Improved Cancer Therapy: Phantom Target Measurements," *Int J Hyperthermia,* 1994, Vol. 10, No. 2, pp. 189-208.

[8] Fenn, A.J. and G.A. King, "Experimental Investigation of an Adaptive Feedback Algorithm for Hot Spot Reduction in Radio-Frequency Phased Array Hyperthermia," *IEEE Trans Biomedical Eng,* Vol. 43, No. 3, 1996, pp. 273-280.

[9]   Fenn, A.J., "Thermodynamic Adaptive Phased Array System for Activating Thermosensitive Liposomes in Targeted Drug Delivery," US Patent No. 5,810,888, September 22, 1998.

[10] Sathiaseelan, V., A.J. Fenn, and A. Taflove, "Recent Advances in External Electromagnetic Hyperthermia," In: Chapter 10 of *Advances in Radiation Treatment*, Mittal, B.B., J.A. Purdy, and K.K. Ang, (eds.), Boston, Massachusetts: Kluwer Academic Publishers, 1998, pp. 213-245.

[11] Fenn, A.J., V. Sathiaseelan, G.A. King, and P.R. Stauffer, "Improved Localization of Energy Deposition in Adaptive Phased Array Hyperthermia Treatment of Cancer," *J Oncol Management*, Vol. 7, No. 2, 1998, pp. 22-29.

[12] Fenn, A.J., B.A. Bornstein, G.K. Svensson, and H.F. Bowman, "Minimally Invasive Monopole Phased Arrays for Hyperthermia Treatment of Breast Carcinomas: Design and Phantom Tests," *International Symposium on Electromagnetic Compatibility*, Sendai, Japan, 17-19 May 1994, pp. 566-569.

[13] Fenn, A.J., A.Y. Cheung, and H. Cao, "Adaptive Focusing Experiments for Minimally Invasive Monopole Phased Arrays in Hyperthermia Treatment of Breast Cancer," *16th Annual IEEE Engineering in Medicine and Biology Society Int Conf*, Baltimore, Maryland: November 3-6, 1994, pp. 766-767.

[14] Fenn, A.J., "Minimally Invasive Monopole Phased Arrays for Hyperthermia Treatment of Breast Cancer," In: *Proc. 1994 Int. Symp. on Antennas*, Nice, France: November 8-10, 1994, pp. 418-421.

[15] Fenn, A.J., "Minimally Invasive Monopole Phased Array Hyperthermia Applicators and Method for Treating Breast Carcinomas," US Patent No. 5,540,737, July 30, 1996.

[16] Fenn, A.J., "Adaptive Focusing Experiments With an Air-Cooled 915-MHz Hyperthermia Phased Array for Deep Heating of Breast Carcinomas," *Proc of the Surgical Applications of Energy Sources Conference*, Estes Park, Colorado, May 17-19, 1996.

[17] Fenn A.J., G.L. Wolf, and R.M. Fogle, "An Adaptive Phased Array for Targeted Heating of Deep Tumors in Intact Breast: Animal Study Results," *Int J Hyperthermia*, Vol. 15, No. 1, 1999, pp. 45-61.

[18] Gavrilov, L.R., J.W. Hand, J.W. Hopewell, and A.J. Fenn, "Pre-clinical Evaluation of a Two-Channel Microwave Hyperthermia System with Adaptive Phase Control in a Large Animal," *Int J Hyperthermia*, Vol. 15, No. 6, 1999, pp. 495-507.

[19] Gardner, R.A., H.I. Vargas, J.B. Block, C.L. Vogel, A.J. Fenn, G.V. Kuehl, and M. Doval, "Focused Microwave Phased Array Thermotherapy for Primary Breast Cancer," *Ann Surg Oncol*, Vol. 9, No. 4, 2002, pp. 326-332.

[20] Vargas, H.I., W.C. Dooley, R.A. Gardner, K.D. Gonzalez, S.H. Heywang-Kobrunner, and A.J. Fenn, "Success of Sentinel Lymph Node Mapping After Breast Cancer Ablation With Focused Microwave Phased Array Thermotherapy," *Am J Surg*, Vol. 186, 2003, pp. 330-332.

[21] Vargas, H.I., W.C. Dooley, R.A. Gardner, K.D. Gonzalez, S.H. Heywang-Kobrunner, and

A.J. Fenn, "Focused Microwave Phased Array Thermotherapy for Ablation of Early-Stage Breast Cancer: Results of Thermal Dose Escalation," *Ann Surg Oncol*, Vol. 11, No. 2, 2004, pp. 139-146.

[22] Fenn, A.J., *Breast Cancer Treatment by Focused Microwave Thermotherapy*, Sudbury, MA: Jones and Bartlett, 2007.

[23] Vargas, H.I., W.C. Dooley, A.J. Fenn, M.B. Tomaselli, and J.K. Harness, "Study of Preoperative Focused Microwave Phased Array Thermotherapy in Combination With Neoadjuvant Anthracycline-Based Chemotherapy for Large Breast Carcinomas," *Cancer Therapy*, Vol. 5, 2007, published online (www.cancer-therapy.org), November 25, 2007, pp. 401-408.

[24] Dooley, W.C., H.I. Vargas, A.J. Fenn, M.B. Tomaselli, and J.K. Harness, "Randomized Study of Preoperative Focused Microwave Phased Array Thermotherapy for Early-Stage Invasive Breast Cancer," *Cancer Therapy,* Vol. 6, 2008, published online (www.cancer-therapy.org), August 25, 2008, pp. 395-408.

[25] Vargas, H.I., W.C. Dooley, A.J. Fenn, M.B. Tomaselli, and J.K. Harness, "A Protocol for Focused Microwave Thermotherapy in Combination with Neoadjuvant Anthracycline-Based Chemotherapy for Investigating Tumor Response of Large Invasive Breast Carcinomas," *American J Clin Oncol*, Abstract, October, 2008.

About the Author

Alan J. Fenn is a senior staff member in the Advanced RF Sensing and Exploitation Group at Lincoln Laboratory, Massachusetts Institute of Technology. He is deputy manager for antenna measurements in the RF Systems Test Facility at the Lincoln Laboratory. He has conducted over 30 years of research in the area of phased array antennas. He joined the Lincoln Laboratory in 1981 and was a member of the Space Radar Technology Group from 1982 to 1991, where his primary research was in adaptive phased array antenna design and testing. From 1992 to 1999 he was an assistant group leader in the Radio Frequency Technology Group, where he managed programs involving measurements of atmospheric effects on satellite communications. From 1978 to 1981, he was a senior engineer in the Antenna Systems Design/Analysis Group in the RF Systems Department at Martin Marietta Aerospace, Denver, Colorado. He received a B.S. from the University of Illinois at Chicago in 1974, and an M.S. in 1976 and a Ph.D. in 1978 from The Ohio State University, Columbus, all in electrical engineering.

Dr. Fenn was elected a fellow of the IEEE in 2000 for his contributions to the theory and practice of adaptive phased array antennas. He was technical program cochairman of the 2001 IEEE Antennas and Propagation Society Symposium. He has served as an associate editor in the area of adaptive antennas for the *IEEE Transactions on Antennas and Propagation*. In 1990 he was a corecipient of the IEEE Antennas and Propagation Society's H.A. Wheeler Applications Prize Paper Award. He also received the IEEE/URSI-sponsored 1994 International Symposium on Antennas (JINA 94) award. In addition to this book, Dr. Fenn is an author of two books and is a coauthor of one book chapter as well as the author of numerous journal articles, patents, short-course lectures, and conference presentations on adaptive phased array antennas and on hyperthermia treatment of cancer.

Index

For further information on these and other Artech House titles, including previously considered out-of-print books now available through our In-Print-Forever® (IPF®) program, contact:

Artech House
685 Canton Street
Norwood, MA 02062
Phone: 781-769-9750
Fax: 781-769-6334
e-mail: artech@artechhouse.com

Artech House
46 Gillingham Street
London SW1V 1AH UK
Phone: +44 (0)20 7596-8750
Fax: +44 (0)20 7630-0166
e-mail: artech-uk@artechhouse.com

Find us on the World Wide Web at: www.artechhouse.com